THE PLANTS
WE NEED TO EAT

By the same author:

A Feast of Oils (co-author)
The Fats We Need to Eat

The PLANTS WE NEED TO EAT

DISCOVER THE POWER OF NATURE'S MIRACLE NUTRIENTS

Jeannette Ewin

Thorsons
An Imprint of HarperCollins*Publishers*

This book is dedicated to those who stand on the frontline of the battle to improve and maintain our good health: the women and men choosing and preparing the food we share around the family table.

Thorsons
An Imprint of HarperCollins*Publishers*
77–85 Fulham Palace Road
Hammersmith, London W6 8JB
1160 Battery Street
San Francisco, California 94111–1213

Published by Thorsons 1997

1 3 5 7 9 10 8 6 4 2

© Jeannette Ewin 1997

Jeannette Ewin asserts the moral right to
be identified as the author of this work

A catalogue record for this book
is available from the British Library

ISBN 0 7225 3278 4

Printed in Great Britain by
Caledonian International Book Manufacturing Ltd,
Glasgow

CONTENTS

Acknowledgements vi

Introduction 1

1 Food! Glorious Food! 12

2 Substance for Life 23

3 A Directory of Foods from Plants 81

4 Questions and Answers about Health 176

5 Tips on Buying and Using Foods from Plants 192

Glossary 206

Sources and Suggested Reading 227

Index 230

ACKNOWLEDGEMENTS

Thank you to the many people who contributed to this book. Special thanks go to the staff of Phytomer, St Malo, France, for information about nutrients in sea plants, and the staff at Gateway Books for helping sort out the kombucha story. A special thanks also goes to Elizabeth Hutchins, who did such a great job of editing the text, and everyone responsible for creating such a fine cover. In particular, however, I am grateful to Wanda Whiteley, Senior Editor at Thorsons, for her insight, patience and support during the long months I spent struggling with this manuscript, and to my husband, Richard, for his help.

JVE

INTRODUCTION

What can give Man hope is that he alone can look back and see what made him, and look forward to gauge what that knowledge implies for his future. Among that knowledge is a fact so simple that it is strange it should so often be ignored: whatever else he may be, Man is a particular structure of organic chemicals. To maintain that structure in good order the materials of which it is built – that is to say food – must be right.

Nutrition and Evolution,
Michael Crawford and David Marsh

Why write another book on food and health?

Bookshelves overflow with thin and fat volumes explaining how to choose foods for good health and long life: why add to the existing heap?

There are three reasons. First, despite the masses of literature about diet and healthy food choices, something remains terribly wrong: we fail to eat the nutrients our bodies need to grow, repair themselves and fight disease. As implausible as it sounds, many of us eat an overabundance of food, but are undernourished. A recent study of typical American eating habits showed they consume significantly more calories than a decade ago, but receive only a portion of the nutrients their bodies need.[1] They are stuffing themselves with high-calorie, low-nutrient foods. Eating habits are similar in the rest of the developed world. As a result the risk of developing a killer disease like cancer and heart disease is on the increase. This will only change when we discover how to appreciate and enjoy a variety of natural, delicious, nutritious foods.

What are nutrients? They are Nature's building blocks, the material from which our bodies are constructed. To grow, to heal and to enjoy a long life, every day we need to eat foods containing a balanced and adequate blend of nutrients.

Which foods are packed with a healthy balance of nutrients? Foods from plants: food prepared from roots, leaves, grains, fruit, stalks and seeds. But many of us find vegetables and fruit uninteresting. Many of us – mostly men – actually avoid them. As a result, our health suffers. What can we do?

We can take action! First we can discover which foods contain miracle nutrients, then understand what the human body needs for health and finally learn how to use this information when shopping for and preparing delicious meals.

Part of our problem with food choices is our dependence on quick-to-prepare, pre-packaged foods. When we are under pressure from work or family stress, many of us reach for ready-mixed, plastic wrapped products instead of the fresh broccoli, cabbage, spinach, apples, dark red berries and wholegrain foods that contain the nutrients we need to calm and restore ourselves. Pre-packaged foods are not good sources of balanced nutrients because most are made with processed flour (stripped of many nutrients), substantial amounts of saturated fats and too much purified sugar. Even some brands of table salt now contain sugar. Foods from plants are excellent alternatives to these unhealthy, processed, high-fat choices. They look great, smell wonderful, are a delight to taste and are also low in damaging saturated fats.

The amount of saturated fat we consume is a well documented cause of serious illness. Unfortunately, many snack foods derive much of their 'mouth-appeal' from this health time-bomb. As a case in point, certain crackers have that melt-in-the-mouth appeal because they are sprayed with an invisible film of a specific form of saturated fat. For good nutrition, a small bag of roasted (unsalted) crunchy nuts is a better choice than a bag of salted crackers or crisps (potato chips).

Eating away from home is another great opportunity to make bad food choices. Have you ever told yourself that that slip of lettuce, two thin slices of tomato and piece of rubbery pickle in a fast-food hamburger will satisfy your daily need for vegetables? I know people who firmly believe it does.

Most of us, however, know the rules of healthy eating and like to think we follow them. For example, we know we should eat at least five portions of fruit and vegetables each day. We know that good medical evidence shows a link between our affluent Western diet and the increasing rates of degenerative illnesses – including heart disease, cancer, and crippling asthma and arthritis. And yet, despite all this, we continue to select foods that fail to give our bodies what they need. We are at war with our appetites and fear the consequences of the food on our forks. We are stuck in the fast-food trap and need help finding healthy alternatives.

This is the second reason this book is important. We need to know more about the delicious foods we *can* eat and relax while enjoying them. To achieve this happy state of affairs, there are some food fallacies that need debunking. For example, many of us believe being 'thin' is the same as being in good health. This is not necessarily true: being grossly overweight has obvious mental and physical consequences, but there is no proof of a direct causal link between being pleasantly plump and developing a heart condition, cancer or any of the other killer diseases that put an upper limit on our life expectancy. Also, in an effort to achieve a thin, desirable image, many of us remove certain foods from our diet (nuts, oils and seeds for example) and dramatically reduce the total amount we eat. We fail to see that eating too little food means consuming too few essential fats, vitamins and minerals. And there is growing evidence that not eating enough of these crucial natural nutrients is linked to developing a life-threatening disease.

A busy account executive I know is certain she eats a healthy diet. See if this sounds familiar. On a typical working day she has one slice of wholemeal toast with low-fat spread, a glass of orange juice and a cup of black coffee for breakfast; coffee mid-morning whitened with a little powdered cream substitute; a 170 gram (6 oz) container of low-fat fruit yogurt and a piece of fruit washed down with a diet cola for lunch; and tea with two biscuits (cookies) to take the edge off her mid-afternoon hunger. Because she lives alone, her evening meal consists of a name-brand, pre-packaged, low-fat chicken curry and rice for one, a salad made with half a tomato and an eighth of a head of iceberg lettuce (dressed only with salt, pepper and lemon juice), a glass of red wine and a scoop of designer ice cream enjoyed as a treat at the end of a long day.

What can we learn from this? First, this young account executive's diet contains too little complex starch: more grains and flour would help build up her intake of B-complex vitamins, for example. Also, although she eats two servings of fruit/juice (the bits of fruit in the yogurt are too small to count), she needs to eat more vegetables, particularly of the green and crunchy type. Although my friend has successfully cut back on fat, especially saturated fat, those little treats with tea and after dinner probably eliminate any benefits she receives from sticking to a low-fat spread and yogurt. (That coffee whitener, by the way, is most probably made with saturated fat. Although the package may say 'vegetable fat', that can mean an oil that has been hardened by hydrogenated or saturated fat from a plant source – coconut oil, for example. The curry may also contain saturated vegetable fats.) Good alternative food choices for my friend would be an exotic fruit salad or half a ripe melon filled with raspberries for her dessert. In terms of calories, these would have 'cost' less than the ice cream. To boost her intake of essential fats, minerals and vitamins, a small container of unsalted roasted nuts would have made a mouth-satisfying afternoon snack. A tasty vinaigrette would add some more essential fats as well as boost the appeal of her salad (which should be about twice the size to provide any real quantity of nutrients). Or, instead of making the salad, if she had taken the same length of time to prepare some mixed vegetables dressed with a well seasoned vinaigrette made with a blend of safflower and walnut oil, the nutritional value of the meal would have been greatly improved.

In summary, the modest number of calories in my friend's diet will probably help maintain her slim waistline, but the foods from which they come will not provide the nutrients needed for the constant rebuilding and repair of her body. Under stress, if she decides to have a baby or during a long illness, her need for basic nutrients from her diet will be even greater, making her existing diet even more unsatisfactory.

Part of the health problem in affluent societies is that we no longer enjoy cooking. The 'art' of cooking – that hands-on, intimate link between ourselves and Nature's own sources of vitality – has become neglected, denied and made to appear banal by quick-fix, pre-packaged foods pushed at us by multinational companies expanding on our gullibility. Food is not just for supplying the stuff of life; it is also for sharing, enjoying with others and taking us back into intimate contact with the world around us. How many of us take time to smell a ripe melon before we cut into it, feast our eyes on the vivid reds, yellows and greens of the fresh fruit and vegetables in our back garden or allow ourselves the satisfaction of mixing, kneading and baking our own bread?

I hope this book gets you cooking and tasting foods you may not have prepared for years. I also want you and your family to try new tastes and textures, enjoying totally new dishes. To set you on your way to discovery, more than 100 foods are described in a directory format that emphasizes their individual culinary characteristics and health benefits *(see pp.81–175)*. I hope this will stimulate you to re-examine natural foods that are truly 'health' foods and discover the sensual pleasure they can give. Included with all of this is a little of the history and fun to be found in food.

The third and final reason for this book is to bring you information about the health benefits from edible plants that could revolutionize the way we approach the prevention of disease. For more than 50 years scientists have excited the medical community by identifying and mapping the biochemical importance of essential substances we know as vitamins; today, that excitement surrounds new knowledge of plant chemicals not essential for the biochemical processes of the human body but able to influence and greatly benefit them. For example, many fruits with black or red skins contain plant chemicals that help control infection. So medical research now confirms that cranberry juice helps control bladder infections, a fact long known by cooks and herbalists in the New England states of America.

Every month something new is published about the healing and protective powers of substances isolated from plants. Organic chemicals in certain common plants – soya beans,

Shiitake mushrooms, cabbage and turnip tops (greens), for example – have been shown to help fight various forms of cancer and other degenerative illnesses. And there is evidence that truly primitive foods – specifically blue-green algae – may become an important source of nutrition. One day, our daily choice of edible plants will not only sustain and pleasure us, but also treat the specific physical conditions which plague us.

Why plants are better than animals

There is confusion about the relative nutritional value of foods from plants and foods from animal sources. Plants and animals do not contain all of the same nutrients because their structures and life cycles are totally different.

Excluding the skeletal structure, food from animals (mostly organs and muscles) is made up almost entirely from protein, but is also rich in saturated fats. Most fats are found just under the skin and in certain parts of the body cavities. Animal flesh also concentrates a few nutrients which are missing or found in very low quantities in plants: vitamin B_{12} is an example. These vitamins play specific roles in the biological processes of animals that are unlike those found in plants. There are non-animal alternatives, but the nutrients are less concentrated. This is why vegans must select their foods with considerable care.

But animals are unable to manufacture energy from sunlight *(see p.25–6 for information on photosynthesis and the origin of nutrients)* and unlike plants, they contain no fibre and no complex carbohydrate (except for the small amounts of glycogen stored in the liver and muscles). By contrast, plants are rich in vitamins, essential fats, minerals, fibre, needed for a healthy digestive system, and sugars and complex carbohydrates, needed for energy. Parts of plants are also good sources of protein.

Unfortunately, even when we eat foods made from plants we may not receive the full measure of their natural goodness. Good advertising campaigns by food manufacturers and fast-food chains have increased our desire for sugar-coated, over-processed and bizarrely shaped breakfast cereals, characterless white bread, mangled meat and synthetic fat. Divested of nutrients and full of things that do us little good, many of these foods are designed in taste and appearance to make us want to come back for more.

Food choices are deeply ingrained in our culture and advertising is only one form of propaganda that affects us. Some influences are very complex. Consider the impact of food rationing during and after World War II. While meat and dairy products – along with flour, sugar and most other basic cooking ingredients – were rationed (although available at outrageous prices on the black market), the vast majority of people ate simple food made from homegrown fruit and vegetables. Looking back with hindsight, they ate a meagre but very healthy diet. Unfortunately, as is often the case, the foods that were rationed and difficult to obtain became desirable. After the war, people wanted to enjoy what they had been denied. Meat was on the menu – and no amount of carrots, leeks, turnips and cabbage could replace it. Nutrient-rich fruits and vegetables were blighted by the stigma of association with hard times. In Great Britain, people were so tired of eating soya protein during the war most are repelled by it to this day!

After the war, encouraged by experts who felt protein was the answer to many health problems, and by a growing and powerful food industry, people lost respect for the edibles that truly contain the essentials of life: grains, leaves, fruit and roots. Many of us are now paying the price. Overweight and nutritionally underfed, we turn to our doctors for easy answers.

'Please give me a pill,' I once heard a woman say to her doctor. Her cholesterol level was sky high. 'I can't follow that diet you gave me, it's all tasteless rabbit food. What is there to life if you can't enjoy what you eat?'

What a pity! By denying herself what she called 'rabbit food', she was denying herself tasty and mouth-filling dishes: oatmeal, rich sauces of tomatoes and garlic on pasta, barley and carrot soup, walnut and raisin cake and the occasional glass of wine would have all helped her fight the cholesterol problem and worked to quell her hunger pangs. Hearty foods are often the most healthy foods.

A healthy body is not a right, it is a blessing and a responsibility. It needs ample materials from which it can maintain and repair itself. It is something we cannot take for granted. When our body fails – which it will – many of us feel that it has let us down, instead of asking what we can do to help it repair itself. We need to rediscover food so we understand the options. We need good food to heal.

Isn't this just another propaganda piece supporting vegetarianism?

No. *The Plants We Need to Eat* does not suggest that you become a vegetarian. That is a matter of individual choice, although, as you will see, all of the protein needed by a healthy adult is easily available in foods from plants if they are properly mixed or balanced.

If this book has a 'political' theme, it is that we should better understand the food we eat. It is a call to exercise our power over food manufacturers and suppliers by asking for and buying items we know will meet our nutritional requirements. This is a proclamation to reclaim ourselves, our bodies and our state of well-being through sensible eating.

How is this book organized?

The Plants We Need to Eat was written to help people get to know non-animal foods again; the challenges this presents are summarized in the first chapter. Chapter 2 explains which nutrients we need and which foods are the best sources of these substances. Chapter 3 provides information on more than 100 of the healthiest foods from plants. Many entries include information on special health benefits and food history; some of these may be surprising. For example, many of the foods we associate with the healthy Mediterranean diet – tomatoes, aubergine and peppers – are not native to that region, but were imported to Europe from the Americas by Christopher Columbus and other early explorers. The Romans, great lovers of food that they were, never saw a tomato or roasted a pepper.

There are practical suggestions as well. Chapter 4 contains questions about foods and specific health problems, and Chapter 5 offers tips on how to make the most from foods derived from plants, how to encourage your children to enjoy finding out about foods, and how eating more fruit, vegetables and foods made with grains can help you get the most value from your food budget.

A LOOK TOWARDS THE FUTURE: WHAT'S IN IT FOR US?

Intensive farming methods, bioengineering and vastly improved distribution systems are not only changing the variety and amounts of food on supermarket shelves, but are also changing the nutrient content of that food. How will this affect our future health? The answer has many parts.

The chemicals used in intensive farming can introduce substances into the food chain which we do not want in our bodies; for example, in Great Britain, cooks are advised to peel carrots to remove the residue of artificial fertilizers and other agricultural chemicals that are deposited in the skin. But peeling can also remove the many nutrients found in the skin of root vegetables. Organic farming is the answer to this form of food pollution, but more Government support is needed for farmers willing to make the shift to this gentler form of growing food.

Exotic fruits and vegetables are shipped around the world by boat and plane. These are often 'cash crops'; not part of the normal diet in a region or country, but saleable to affluent buyers. As a consequence, land needed to support local needs is diverted into money-raising ventures, often throwing the local nutritional and economic life out of balance. What is more, the nutritive value of the food shipped may not be as high as that enjoyed in its native environment: fruits are often picked before they are ripe, leaving nutrients unformed; long periods between the harvest and sale of a food can cause an ageing and loss of nutrients. Food richest in nutrients are often those grown closest to home.

In the future, bioengineering and genetic selection will certainly improve many plants. Researchers at the University of North Dakota have produced a new soya bean which is about 90 per cent protein and no fat.[2] Knowing the potential importance of soya protein in controlling medical conditions as different as menopause symptoms and prostate cancer *(see pp.94–6)*, this new breed of soya could help supply millions of people around the world with food that can significantly improve their health.

But all new developments may not be ideal. Another group of scientists improved the quality of soya protein by implanting it with genetic material from Brazil nuts; unfortunately they also transferred material containing a gene which introduced an allergic factor.[3] Before scientists can chop and change genetic material and produce more perfect foods, they need to be cautious and respectful of what Nature provided in the first place.

Finally, I believe the near future holds a time when medical practitioners will agree that food choices should be planned and determined by changes in our physical health. By this I do not mean simply giving advice about cutting back on saturated fats or eating more fibre, as happens today. I see a time when bladder infections will automatically be treated with antibiotics in combination with servings of cranberry juice, when the symptoms of menopause will be controlled by adding tasty, phytoestrogen-rich foods made with soya protein and yams to a woman's diet, and when we cook with green tea and Shiitake mushrooms to strengthen our immune systems.

In the future, more than ever before, what we eat will determine how long we live and how much good health we enjoy during those years, and the wise and creative use of edible plants holds the secrets of maximum well-being.

One final thought:

Food is about history, food is about life and food is about enjoyment.

Notes

1 *Nutrition Reviews* 54 (1996) no.2, pp.68–9
2 *Gene Therapy Weekly*, 25 March 1995, p.8
3 *New England Journal of Medicine*, 13 March 1996

The suggestions and ideas in this book are in no way meant to replace the advice and services of trained health professionals. Any concerns regarding your health or well-being should be discussed with your doctor. Suggestions presented here do not carry any claims for prevention or cure and are to be tried only at the reader's discretion.

1

FOOD! GLORIOUS FOOD!
A FRESH LOOK AT AN OLD SUBJECT

The Lord hath created medicines out of the earth; and he that is
wise will not abhor them.

Ecclesiasticus XXXVIII

GREEN MIRACLES

How much suffering could be avoided and money saved if medical research were to invent a single substance that could lower
rates of breast, colon, lung and prostate cancer, reduce cholesterol levels, fight free radicals, improve immunity against infection and reduce the distress of the menopause! All the world
would be grateful.

But what if I said there was a simple food – a bean not much
bigger than your little fingernail – that contains a rich supply of
natural plant chemicals (phytochemicals) able to perform all of
these medical miracles and, at the same time, provide a rich
source of healthy protein low in saturated fat? And what if I
cooked up a plate of these beans and offered them to you for
dinner? On their own soya beans are not very attractive – rather
pale looking and lacking in some of the flavour and texture
needed to impress gourmet cooks. My bet is that 90 per cent of
people offered a meal of plain soya beans would take a quick
taste, shrug their shoulders and pass on by. However, prepared
in an interesting way, and in their many forms, soya beans and
soya protein make delicious soups, main courses and creamy
desserts.

Grossly undervalued, the soya bean is an excellent example

of the plants we need to eat but often reject as part of our diet. And what about cabbage, broccoli, Brussels sprouts and all of the plants in the mustard, or *Cruciferae*, family we now know contain phytochemicals that fight cancer? Broccoli suffered rejection at the highest level when US President George Bush said it was no longer welcome at the White House. What a pity. Perhaps some future First Family will set a national example by giving the culinary pride of place to broccoli, or generous servings of one of its many close relatives, on every state occasion menu. Over the years, by helping to popularize this important food, they would make a significant contribution to reducing the risk of cancer in the United States.

There are about 1,900 species of *Cruciferae* growing in temperate climates around the world. As the name suggests, they all have flowers with four petals that take the form of a cross. Some are weeds, some are grown for cattle food, some are valued edible plants. In parts of Europe, up to 30 per cent of land cultivated for vegetables is used to grow *Cruciferae*. *Brassicas*, a genus within this family, have recently gained special recognition among medical experts because they are especially rich in cancer-fighting substances. Cabbage, cauliflower, kale, swede, turnips and, of course, President Bush's broccoli are among the *Brassica* that have been grown in Europe for centuries as food. They do have a strong smell when cooked and this reduces their appeal. Nevertheless, good cooks can find ways to solve this problem and, if we include at least one vegetable from this group in our diet each day, we are protecting our bodies from major killer diseases.

What are some of the other plants that contain miracle nutrients?

* Cranberry juice contains a substance that forms a protective film over the lining of the bladder; on its own, this natural plant substance repels bacterial growth and helps control the pain and discomfort of cystitis. Working with a prescribed antibiotic, cranberry juice speeds healing and reduces the amount of medication required.
* Green tea contains a substance that kills *Streptococcus mutans*, a bacterium known to cause tooth decay.

* Lentinan, a substance found in Shiitake mushrooms, a delicious ingredient in Asian food, has been shown to stimulate the immune system and fight the growth of cancer cells. In Japan, lentinan is used as a cancer treatment.
* Spirulina, blue-green algae, is an example of an aquaherbal micro-organism with outstanding nutritive and healing properties. Experts believe this primitive food has been eaten by various groups for more than 1,000 years. High in protein, essential fatty acids, beta-carotene and vitamin B_{12}, Spirulina contains many of the essential nutrients that are known to lower blood fats and protect against damage from free radicals. It also contains phytochemicals that stimulate the immune system. Along with Chlorella, another algae, and various forms of seaweed, Spirulina may be a food of the future.

The list of plants containing miracle substances (sometimes called 'associate nutrients') is long, and Chapter 3 provides information about individual foods and what they contain. Here, it is enough to recognize that we will live longer, healthier lives if we enjoy more and a wider variety of plants as food.

FOOD, ILLNESS AND A TIME FOR CHANGE

In 1946, Dr Hugh Sinclair, a leading British expert in nutrition, felt strongly about the importance of diet to health and sought support for a permanent department of nutrition at Oxford University. According to Sinclair, the idea was cast aside by the university, although some members of the academic community had previously given the concept provisional acceptance. Funding was available to support such a department, but there was no enthusiasm for it within the academic community; allegedly the reason given was that nutrition was at that time a subject close to exhaustion. Work done during the previous three decades had demonstrated there was a finite number of chemical substances needed for the human body to function properly, and so the best scientific minds at Oxford – many of them Nobel Laureates – agreed that a university department set aside for the specific study of nutrition would soon become a white elephant and drain the university of desperately needed resources.

These experts ignored the fact that identifying the ingredients that make up the stuff of human existence, along with understanding their basic interactions, is only part of the story. When playing the piano, learning scales and fingering are fundamental, but melody, key, chords, rhythm and shading of touch are needed to change simple notes into music. So it is with nutrition: knowing the nutrients is not enough. Research is also needed to establish how these substances are influenced by genetics, environmental factors, the balance of nutrients consumed, illness and time.

Today, the list of essential nutrients needed for human health and well-being has changed little from that committed to memory by every medical student 50 years ago. Where we now have an advantage over the past is in our understanding of how nutrients work together. We now know that the demand for specific nutrients is not static, but affected by stress, environmental pollution, infection, injury – the list goes on and on. In other words, we are coming to understand that health can be better sustained by altering the balance of specific nutrients in our diet as our life circumstances change. What we must now learn is how this information is reflected in the foods we choose to eat.

Hugh Sinclair argued that most modern illness is caused by dietary deficiencies created by our Western diet. Most concerning were the effects of a special group of polyunsaturated fats known as 'essential fatty acids'. Essential fatty acid deficiencies take time to develop and their symptoms are diverse. Like other vitamins and essential nutrients, deficiency probably first affects a single genetically susceptible organ system of the body – the heart, skin or immune system for example – and progresses from there. As the symptoms and their severity vary from person to person, it is difficult to define a specific syndrome or set of conditions common to all people experiencing the deficiency.

The possible role of future medical practitioners will be to help each individual achieve a balanced, individual chemistry. A doctor – of some sort – will perform a nutritional assessment on us when we are ill, before even listening to our symptoms.

Then, in a perfect world, we will be offered a choice of nutritional means by which we can adjust our inner balance before we are forced to accept artificial (although many times highly effective) medical substances.

Until then we must work to change our culture ... and our food preferences.

THE DIET WE CHOOSE

A well-known writer ... wrote a book about honey. The statement in that book – 'If you were isolated on a desert island, honey is the one thing you could live on indefinitely' – would make an intriguing subject for a Final Year degree examination reworded: 'If you eat nothing but honey what nutrient deficiency would you die of?'

Dr Hugh Sinclair, in *Applied Human Nutrition for Food Scientists and Home Economists*

Balance among the foods in a diet is fundamental for good health. There are no magic mixtures, no simple formulas. Balance can only be achieved by eating a little bit of many things. We cannot gulp down a prepared 'health' drink or munch a box of 'diet' biscuits and expect that to take care of all of our problems. We must eat food; whole food. And in the right quantities and combinations.

It is not enough to *say* you are going to eat a balanced food intake – you must be ready to act on that statement and make a long-term effort to achieve the desired results. That is harder than it sounds. All too often we are not aware of what is in the foods we eat. That means we need to learn before we choose. Take the facts about the amount of saturated fat we eat as an example.

When early humans were hunter-gatherers, and later when workers in agrarian communities were without the assistance of motors and electricity, animal fat (most of which is saturated) provided a rich source of calories for long journeys by foot and stored energy for heavy work. Hard and long lasting, certain fats – especially the hard, saturated fat from beef cattle – were also used in making candles to light living-rooms and churches, in

THE PLANTS WE NEED TO EAT

polishing boots and waterproofing tough fabric. In the scheme of life, animal fat had its place both in the diet and in providing daily comforts. But things have changed.

In their book *Nutrition and Evolution* (Keats Publishing Inc., 1995), Michael Crawford and David Marsh point out that with the advent of modern labour-saving devices, motor cars and sedentary lifestyles, we no longer need large quantities of high-energy fat in our food. We don't move around enough to burn off the extra calories it adds to our diet. But instead of reducing the total fat, particularly the saturated fats, we are consuming more! Saturated fat is a common ingredient in modern snack-foods, ice-cream substitutes, sausages, pies, quick-fix foods and butter substitutes.

To illustrate the problem, Crawford and Marsh calculated that the energy stored in the fat of meat consumed in the United Kingdom each year represents about '1.097 times 10 to the power of 14 Joules a day, or enough to keep an oil-fired 1200 megawatt power station in operation for a year'. That is enough electricity to light all the electric bulbs in all the homes around the country. But we don't use the fat to generate electricity; nuclear energy or fossil fuels are used. Instead, Crawford and Marsh conclude, 'We now eat the candles.'

As our tastes adjust to enjoy more and more fatty foods, we worry more and more about our waistlines. This apprehension supersedes our concern about eating adequate quantities of nutrients. As already mentioned, while trying to eliminate unwanted calories, many of us avoid eating sufficient quantities of food rich in polyunsaturated fats, such as salad oils, nuts and seeds. We also shun bulky starchy foods – including pasta, cereals and potatoes – and blame them for our expanding girth. In fact, these are exactly the foods we need to eat, and in large quantities, to ensure we obtain adequate supplies of essential nutrients. And we are what we eat. If the basic building blocks of the body are not in place, or are in some way flawed by the lack of essential components, we cannot function properly and are prey to illness and premature death.

A child born in affluent cultures today can expect to live almost twice as long as a child born 150 years ago because we have learned how to cure and control the infections and conditions that once swept away the very young. However, we have not achieved the same success in prolonging life after a person reaches middle age.

Hugh Sinclair calculated that in 1841 a man of 50 could expect to live another 20 years, to the age of 70, or the Biblical 'three score years and ten'. Today, a man of the same age can expect about the same fate – to live another 23 or 24 years. That additional three or four years of life expectancy is not much improvement in light of what we have learned about treating illness and disease during the intervening 150 years! Sinclair wrote:

> The year 1841 was before all the great advances in medicine, before anaesthesia, before the use of antiseptics and almost all the drugs we use today. Despite all these advances we cannot do much more to keep a middle-aged man alive today than in the Dark Ages – and that is an astounding fact.

What has gone wrong? What can we do to slow and reverse this rush towards untimely death?

As you may expect by now, the answer rests in the foods we eat. We need to clear most of the processed foods out of our cupboards and refrigerators, slash the amount of saturated animal fat we eat to a small percentage of our current intake and rediscover plants as food. We also need to increase our activity level, but most significantly we need to greatly increase the amount of fruit, vegetables, nuts, seeds, grains and plant-derived foods in our diet.

UNCHANGING ATTITUDES

Concern about the lack of fruit and vegetables in the diet is not new. In 1611, the life of Giacomo Castelvetro, an Italian translator and editor, was saved from the Inquisition by the British Ambassador in Venice, Sir Dudley Carleton. Three years later, while living in exile amid the comforts and generosity of Sir Adam Newton, at Eltham, England, Castelvetro found the heavy British diet of meat and sweets unpleasant and unhealthy.

He longed for the gardens and profuse varieties of foods left behind in his beloved Italy, and three years after his arrival in England he wrote *The Fruits, Herbs and Vegetables of Italy*. This sparkling account of life and food in Italy was meant to encourage the British to eat more fruit and vegetables. Its message was simple: take advantage of the many varieties of edible plants already growing in British gardens for their botanical beauty, and enjoy them also for their delicious taste and health-giving properties.

Copies of the manuscript were distributed to a number of prominent figures, including Lucy Russell, Countess of Bedford, herself a devoted gardener. Castelvetro hoped his descriptions of various garden delights and how they can be prepared and cooked to best capture their taste and texture would encourage her to take up his cause. Unfortunately, the Countess was one of those gardeners more interested in plants for show than for food. No one took the manuscript seriously and the frail escapee from the agonies of the Inquisition suffered instead the agony of seeing his work cast aside and unpublished. At his death, a few years later, there was no evidence his work had influenced the foods and kitchen gardens of his adopted country.

The first English translation of *The Fruit, Herbs and Vegetables of Italy* was published in 1989. Although more than 150 years old, its message could have been written by the World Heath Organization, the American Medical Association, the British Department of Health or any of the other distinguished health organizations of today. And few attitudes have changed since Castelvetro first spoke up for edible plants in seventeenth-century England. Medical experts and Government agencies may wag their fingers, present us with appetizing pictures of fresh foods and warn us to change our ways, but they have done little to actually change our attitudes. Reasons for this are complex, but money and politics play a role. Food producers, including animal farmers, wield considerable political clout. The cry to pay more attention to foods from plants is often carried away on the wind.

There are those, however, who attempt change. As an example, in Britain in 1995, the Honourable Tony Banks (Labour,

Newham North-West) presented a Bill in the House of Commons 'to establish a national body to promote the health advantages of vegetarianism'.[1] Specifically, he called for a fruit and vegetable commission funded by industry and by Government to encourage the growth of vegetarianism. On this occasion, the Bill failed, but not because the case presented in its favour lacked merit.

In support of his Bill, Mr Banks warned that ignorance about our bodies is the greatest killer in our society. Significant quantities of statistical data in the United States of America and from the World Health Organization point specifically to vegetarian diets, or diets containing very little meat, not only saving lives, but also reducing the cost of medical care throughout an individual's lifetime. According to Mr Banks, 'The total annual direct medical cost savings from avoiding meat and tobacco are estimated in the States to be as high as $80 billion a year.' He further said that studies show vegetarians suffer 30 per cent less heart disease and 40 per cent less cancer than those who avoid a plant-based diet. 'Not only do veggies live longer, they look better. Vegetarians are 10 per cent leaner than omnivores,' he stated.

And then Mr Banks turned to the environmental problems associated with meat production. In addition to water and air pollution caused by slurry, sewage and the creation of methane gas by cattle, he also noted that more fossil fuels are used in meat production. Finally, he raised the question of agricultural efficiency:

> Ten kg of vegetable protein fed to livestock will supply only 1 kg of meat, and while meat protein produced on 10 hectares of land will feed only two people, soya protein grown on the same area would feed 61 people.
>
> Millions of people all over the world are dying of starvation, while the world feeds 38 per cent of all crops to animals rather than to human beings ... Add to that the impact of over-grazing, which produces deserts and deforestation, and the price of meat becomes too high for the world to pay.

THE PLANTS WE NEED TO EAT

Despite this, the Bill had little support within Parliament. Perhaps it failed because too many people think vegetarianism is a bit of a laugh. Perhaps because it was offered by a member of the Opposition to Her Majesty's Government of the day. Perhaps most people have not taken the time to learn the basic facts of health and nutrition – or apply them.

In January 1996 the United States Department of Agriculture released the first tranche of data from a new nationwide survey of eating habits among Americans. Research during the 1970s showed the average American consuming 40 per cent of his or her calories in the form of fat. In 1993, that level dropped to 34 per cent, although the total caloric intake was actually higher; a year later, the fat intake dropped to 33 per cent and researchers expect it to fall further towards the perceived ideal of 30 per cent.[2] The bad news is that Americans weigh more today than they did two decades ago. Statistics then indicated one in five adult Americans were overweight; today the number is closer to one in three.

According to Karl Stauber, Undersecretary of Agriculture for Research, Education and Economics, Americans continue to consume low levels of green and yellow vegetables, despite major nationwide campaigns to increase their consumption.[3] Probably as a direct result, American adults do not consume enough zinc or magnesium, and calcium, iron, vitamin E and vitamin B_6 levels in women were below RDA (Recommended Daily Allowances, see p.40–41) levels.

The survey showed that consumption of grains and mixed-grain foods is on the rise: pizza consumption is up over 110 per cent in 20 years. This may mean more people are getting more benefit from part of a Mediterranean diet. But consumption of salty snack foods – like crisps (potato chips), crackers and pretzels – is up more than 200 per cent. As levels of sodium affect blood pressure, the benefits from eating more grain products may be wiped out by eating too much salt.

And so, even in a country where good food is plentiful and cheaper than almost anywhere else on Earth, and where health education and diet are the subjects of national campaigns by the Government and the public media, people still fail to provide their bodies with the nutrients they require.

Understanding basic nutrition takes a bit of effort. But if you find out the facts and try applying them to your own diet and lifestyle, the positive changes in your health and appearance will be convincing.

Notes

1 *Hansard*, 8 March 1995, pp.344–8
2 *Nutrition Reviews* 54 (1996), no.2, p.68
3 Ibid

2

SUBSTANCE FOR LIFE
WHAT FOOD CONTAINS

It is likely that the increase in coronary heart disease and other conditions has been accentuated by the reduced intake of the health-promoting micro-nutrient-rich fibre foods just when there was an increased need for them due to the rise in consumption of fats and free sugars. If so, the advice needed is positive and simple – increase the consumption of crunchy, munchy, chewy wholefood. With a doubling of the nation's intake and variety of fruit, salad, vegetables, pulses, nuts and seeds, and having more unrefined cereals, there would be less room left in the shopping basket for the lighter energy, processed foods.

Sir Francis Avery Jones *Journal of Nutritional Medicine* 4 (1994), pp.99–113

Experts tell us almost everything we eat should come from 'crunchy, munchy, chewy' wholefoods. But what exactly are wholefoods? They are the edible parts of plants that arrive on our tables in forms as close to their natural condition as possible. They are the piece of fresh fruit you eat at lunchtime, and the delicious breads and cakes made from flour still brimming with the vitamins and essential minerals contained in the unmilled grain. Wholemeal foods also include those snacks of dried fruits and nuts that are untreated with excessive quantities of salt and preservatives. They are the roots, fruits, stems and leaves that come to our table free of added saturated fats and artificial colouring used in many manufactured and processed foods that fill our supermarket shelves. They are the rich, natural, balanced source of the nutrients that build, renew and heal our bodies.

Foods made from grains, like pasta, breakfast cereals and bread, should provide about 40 per cent of the calories we eat. Another 30 per cent should come from fresh fruit and vegetables, and an additional 10 to 20 per cent should be derived from nuts, seeds and vegetable oils. High-protein foods should make up a small proportion of our diet; a normal adult only needs about 57 grams (2 oz) of pure protein a day. This can come from animal sources or from beans, grains and other parts of plants.

WHAT ARE THE ORIGINS OF NUTRIENTS?

Nutrients originated in the same place as life itself – in sea water warmed by energy from the sun. This makes sense when you think that all forms of life – plants, fish and so on – are nothing more than highly organized structures built from trillions of nutrient molecules. Acknowledging the risks of oversimplification, the story goes something like this ...

We can only speculate about the exact processes, but scientists believe the basic molecules of life, including the fundamental units of the material which contains the genetic code, or DNA, were formed in the seas over billions of years. They resulted from the combined effects of the bombardment of radiation from outer space on the water of the sea, the presence of minerals dissolved from the rocks of the seabed, and lightning and other external forces of this planet. The elements of the Earth combined and took on positive or negative electrical charges. Atoms and small molecules with opposite charges grabbed hold of each other and formed more complex units. Amino acids, nucleic acids, fatty acids – all of the basic building blocks of life – were probably floating in that same primeval 'soup'. Slowly the structures became more elaborate. Amino acid units joined and formed protein chains. Nucleic acids combined and developed into the double helix of the genetic code. Still other complex structures were formed: proteins and fats fused into molecules with new properties; minerals fitted into the growing structures of proteins.

Slowly, these primordial forms of life, these essential nutrients, merged into more elaborate structures capable of multiplying themselves and consuming and using simpler molecules as spare parts for reproduction. As this occurred, the patterns of biochemical activity were locked into an expanding variety of genetic material. Small protein units (enzymes), prescribed by this evolving genetic store of biological blueprints, became messengers that controlled the building processes and organization of molecules in these emerging lifeforms. Barriers, or membranes, of proteins and fatty acids began forming around groups of active molecules. Through these membranes flowed two types of substances: inwards came those needed for chemical operations within the cell and outwards went the waste products resulting from those same chemical activities.

The great breakthrough in the evolution of life came with the formation of simple one-celled plants – ancient ancestors of the algae we know today. These rudimentary life-forms contained a miraculous molecule: chlorophyll! Next to the development of the genetic code, the evolution of chlorophyll is arguably the most important event in the miracle of life. This complex green molecule absorbs energy from sunlight and, through a series of chemical processes, combines that energy with molecules of water and carbon dioxide to form glucose, a sugar. So light energy is transformed into chemical energy. During this process plants clean the air of the carbon dioxide we rootless forms (all the animals on Earth) produce as waste. In return they supply fresh oxygen needed for animals to survive. Other chemical processes in the plant then take over and molecules of glucose are joined into larger and more complex nutrients called starch. Energy from the centre of our solar system is now our food.

Without plants, life is finite. If all the animals in the world suddenly disappeared, we could survive indefinitely because green plants would continue to generate chemical energy through photosynthesis and store it in forms we could enjoy as food. If, however, all plants died out on the Earth, we and other animal species would survive only as long as there were other animals to eat.

The chemical processes of life are amazingly consistent. The cycle of chemical steps required to release energy from a molecule of glucose is exactly the same in a single-celled fungus, or yeast, as it is in your next-door neighbour's cat. The same basic materials, called 'substrate', are involved, the same enzymes, or control molecules, and the same vitamins and minerals that work as catalysts and 'co-enzymes'. This exact parallel of chemical activity between living things holds for almost all basic life processes. That is why a surprising amount of the genetic code found in your own body is identical to that found in a yeast cell, a slug and a blade of grass.

A major difference between you, the yeast, the slug and the grass, of course, is that the grass contains chlorophyll. Grass can also make its own vitamins, essential fats, carbohydrates and basic protein units, the amino acids. Along with the sugar it manufactures, combined with water and minerals from the soil in which it grows, grass needs little else. With these it can produce the vitamins and the starches, the proteins and the fats it needs to survive, grow and produce the seeds by which it reproduces itself.

Because a blade of grass – or any other green leafy plant – is a powerhouse of chemical activity, it contains a store of the energy and building blocks it needs for its own life processes. Along with amino acids, sugars and fats, there are stores of vitamin C, most of the B vitamins and a balance of minerals. Throughout all of the parts of a plant the fundamental substances our bodies need for life and health are stored or are being used. These substances are Nature's miracle nutrients and are made available to us when we enjoy plants as food.

Plants not only supply our basic nutritional requirements, they also provide unique plant substances, the phytochemicals or associate nutrients already mentioned. These have no role in the set pattern of biological activity in the human body, but they can affect these activities and protect them from extraneous conditions that could alter them. For example, recent scientific studies suggest that oestrogen-like molecules in plants can block the cancer-causing substances that may cause breast cancer. Other types of molecules appear to help repair a damaged heart; still others help lower the level of dangerous cholesterol in blood. Chapter 3 contains specific information about plants that heal *(see pp.81–175)*.

A word about minerals

The minerals in a plant are present as part of its own metabolic purposes. They are therefore present in the combination and balance best suited for these life processes. Enjoying a variety of foods derived from plants provides a healthy mixture of minerals needed for good health. Unhealthy amounts of a single mineral, sodium for example, are not possible. The prolonged excessive intake of sodium – found as part of sodium chloride, or common table salt – has been linked with high blood pressure and implicated as a cause of stroke. Processed foods are often loaded with sodium to add taste or – as salt – act as a preservative. These same levels could not exist in living plants because the quantity of sodium would destroy their normal biochemical balance. *(More is said about minerals on pp.67–80.)*

PLANTS AND ANIMALS

Animals' nutritional requirements

Despite the fact that plants and animals share a surprising amount of genetic material, a number of basic biological processes and many requirements for nutrients to carry out the chemistry of life, they are also amazingly different. Animals must seek their nutrients from sources outside themselves and therefore must be able to move. Movement requires the development of muscle tissue. Muscle tissue is protein. To build protein, animals need a complete source of amino acids.

In general terms, plants can be an excellent source of protein. For the sake of comparison, 100 grams (3.5 oz) of raw chicken contains approximately 21 grams (0.7 oz) of protein, while 100 grams of low-fat soya flour contain almost 45.3 grams (1.6 oz) of protein. The same weight of wholemeal wheat flour contains about 13 grams (0.5 oz) of protein. However, plant proteins differ and they do not all contain the same group of amino acids. This is important in the study of human nutrition, because there are nine amino acids we must receive from our diet because they are not manufactured by our bodies.

The specific types and balance of amino acids manufactured by a plant species are determined by its specific genetic code. Soya protein contains all of the amino acids needed for human health, for example, but rice and beans do not. However, these foods are missing different amino acids; alone, neither provide all we need for good health, but combined as hot and spicy chilli beans served on rice, a healthy balance of proteins becomes an unnoticed 'side benefit' from a delicious meal. Vegans, who eat no form of animal protein, should pay particular attention to the way they select and prepare foods from plants as their source of protein.

Research in the United States showed that '...except for premature infants, soy protein can serve as a sole protein source in the human body'.[1]

Fats are also an important part of animal nutrition: they are needed as a rich energy source, as structural components in cell membranes and as critical components in the 'messenger' molecules that regulate biochemical processes. Animals also need fat as insulation under their skin, and as padding to protect and support certain internal organs. The majority of fat in animals is stored energy. Although species differ, most animal fat is saturated. That is the type of fat doctors believe is dangerous for your heart when eaten in large quantities.

Plants also produce fats. These also form part of cell membranes, take a role in regulating biochemical activity in cells and serve as dense energy stores. However, with only a few exceptions (such as coconut oil) the fats found in plants differ from that found in animals because they are 'unsaturated'. That means each molecule contains exact locations where biological activity or molecular 'bending' can take place. These 'bendable' or 'flexible' locations make unsaturated fats liquids in low temperatures. Saturated animal fats are rigid and are solid in the cold *(for further information, see pp.44–5 and 165–6)*. For now, it is enough to say that plants produce two types of fats that are absolutely essential for good health: omega-6 and omega-3 essential fatty acids. Simple sea plants like blue-green algae are rich sources of these fats and are the reason fatty fish feeding on algae contain high levels of omega-3 fats in their tissues.

THE PLANTS WE NEED TO EAT

Some chemical processes in humans require the presence of certain vitamins not found in plants but produced by certain bacteria and yeasts. Some of these micro-organisms live in the gut of animals, from where the vitamin is absorbed. Vitamin B_{12} is an example. Without this vitamin humans cannot function; that is why people selecting a diet free from all animal protein must make certain they fulfil their nutritional requirements in some other way.

Plants and animals as food

The structural differences of these two living forms are very great. As a result the chemical composition of their structure is different and when used as food, they provide a different balance of nutrients and other substances.

Animal tissue is rich in protein, saturated fat, some vitamins and minerals. Its stores of carbohydrate (glycogen, stored in muscles) are limited and do not include any fibrous form needed for normal bowel function. Through the food chain or intensive farming methods, animals used for food are frequently exposed to antibiotics, hormones, pesticides, fertilizers and abnormal sources of nutrients. Sometimes in farming, like everything else tinkered with by man, things take a dangerous turn in the name of progress. The results can be disgusting: feeding chickens pellets containing their own faeces, and giving cows – ruminants built to live on grass – the desiccated and pelleted brains from dead sheep. (These practices have now ended, but they will continue to haunt us for some time.)

By contrast, plants are a safer source of nutrition. It is true that intensive farming methods can cause contamination by fertilizers and pesticides, but these are easier to spot and control (by washing and peeling) than contamination in meat and animal products. By choosing a mix of foods created from the leaves, roots, bulbs, seeds, fruit, nuts and grains, we can enjoy truly delicious meals that are healthy and sumptuous.

Whenever possible, whether you are eating plant or animal products, opt for organic products.

NUTRITION FROM PLANTS

Our focus here is on true plants: those with roots, stems and leaves (containing chlorophyll), and which produce seeds, nuts or grains that sprout and produce the next generation of plant. Varieties of plants differ because their genetic materials differ – in one case a flower may be large and pink, whereas in a plant that appears almost identical the blossoms may be white and half the size. When genetic material differs, the chemical processes differ, and the exact blend of nutrients needed to fuel and control these processes also differs. For example, the seed oil from one genetic variety of evening primrose plant may contain much more gamma-linolenic acid (an essential fatty acid) than another variety that appears to be identical. As a more extreme example, broccoli and cauliflower – both members of the genus *Brassica* – look different, taste different and contain different amounts of the same substances: broccoli contains approximately 11 times more carotene than cauliflower, and about a third more folate.

Also, different parts of plants contain different substances, or nutrients, depending on their purpose:

* *Leaves*, rich in green chlorophyll, are a good source of minerals, vitamins and phytochemicals used by the plant to turn water, carbon-dioxide and captured sunlight into sugar. They also contain essential fatty acids.
* *Cellulose*, the indigestible form of complex carbohydrate we know as 'fibre', is mainly found in the parts of the plant that give it structure: the stems and stalk, and outer husks of seeds. Fibre is important for a healthy digestive system and some fibre is also thought to help lower blood cholesterol levels.
* *Fruits* – which are meant to fall and rot where new plants will begin and grow – are good sources of carbohydrate, vitamins C, A and B complex, and the minerals iron and calcium. They are rich in the natural 'antioxidants' thought to help fight cancer, heart disease and other degenerative diseases. Vitamin C, one of these antioxidants, also plays a key role in fighting infection. Most fruits are good sources of

potassium, which is needed to maintain normal blood pressure, and many also provide pure water.

* *Tuberous and swollen roots* (potatoes and carrots are examples) store up the starchy energy (carbohydrate) the next generation of plant will need to find in the ground to support its rapid growth. This can also be a good source of protein and vitamin C.

* *Seeds* (including grains and nuts) are rich in starch, protein, unsaturated fats, fibre, minerals (including iron and calcium), vitamin E and vitamins B complex. Often avoided by dieters because of their high fat content, these are the most valuable source of plant micro-nutrients.

The miracle of seeds

Seeds, grains and nuts are the result of fertilization of a plant's flower, or reproductive system. As a result they contain a tiny plant embryo, or 'germ', surrounded by a dense mix of high-energy carbohydrates and oils packaged with minerals and vitamins. Genetically programmed, tiny cells in the germ burst into a rapid-fire series of biochemical processes as soon as the right combinations of temperature and moisture are present. Around this minute powerhouse of activity the store of nutrients fuels the life-fires when they erupt. Once the germ begins to expand and form a sprout, pushing out from the seed and into the surrounding environment, the energy molecules stored in the seed are absorbed and transformed into new molecules that support other processes of life. The first, distinctive leaves of the sprout form and, when exposed to light, the first signs of chlorophyll appear. The changes are remarkable, and are the origin of certain vitamins and phytochemicals.

Compare two forms of the tiny green mung bean: one is dried, then boiled and prepared as part of a dish for dinner. The second is the seed left in a damp, warm place and allowed to sprout. Based on data published in McCance and Widdowson's *The Composition of Food* (Royal Society of Chemistry and Ministry of Agriculture, Fisheries and Food, 1991), we can calculate that 100 grams (3.5 oz) of bean sprouts contain about 90 grams (3.2 oz) of water; the cooked beans contain about 70 grams (2.5 oz) of water.

Table I: Major nutrients in selected foods (100 gram (3.5 oz) portions)

Food	Calories	Protein	Carbo-hydrate
	kcal	gr	gr
Apple, raw	35	0.3	8.9
Banana	95	1.2	23.2
Figs,dried	227	3.6	52.9
Spinach, boiled	19	2.2	0.8
Quorn*	86	11.8	2.0
Potatoes, baked (whole)	136	3.9	31.7
Onions, raw	36	1.2	7.9
Brazil nuts, unroasted	682	14.1	3.1
Walnuts	688	14.7	3.3
Brown rice, boiled	141	2.6	32.1
Oatmeal, quick cook, raw	375	11.2	66.0
Olive oil	899	tr	0
Parsley, fresh	34	3.0	2.7

*Quorn is a high protein food produced from fungus and used as a meat substitute.

Comparing equal weights of the dried material from these two types of bean tells us many things. First, the mineral content of the sprouts demonstrates that it takes up minerals from the water absorbed during formation of the delicate, crisp, rapidly expanding root; chlorine triples, for example. Second, the growing germ uses up much of the fat stored in the bean; the amount drops by two thirds, although the energy stored as carbohydrate changes little during this early stage of plant development. The fats (mainly polyunsaturated essential fatty acids) are used for building cell membranes and complex vitamins needed to energize and control the next burst of growth activity. As evidence, the amount of substances called 'carotenes' in the raw sprouts is more than 10 fold that measurable in cooked mung beans. (These new substances include beta-carotene, the fat-soluble

THE PLANTS WE NEED TO EAT

Fat, total	Fat, saturated	Fat, monoun- saturated	Fat, polyun- saturated	Water
gr	gr	gr	gr	gr
0.1	tr	tr	0.1	87.7
0.3	0.1	tr	0.1	75.1
1.6	N	N	N	16.8
0.8	0.1	0.1	0.5	91.8
3.5	0.6	0.7	1.3	75.0
0.2	tr	tr	0.1	62.6
0.2	tr	tr	0.1	89.0
68.2	16.4	25.8	23	2.8
68.5	5.6	12.4	47.5	2.8
1.1	0.3	0.3	0.4	66.0
9.2	1.6	3.3	3.7	8.2
99.9	14.0	69.7	11.2	tr
1.3	N	N	N	83.1

From *The Composition* of Food, fifth edition, reproduced with the permission of The Royal Society of Chemistry and the Controller of Her Majesty's Stationary Office.

precursor of vitamin A.) Folate increases five fold and thiamin more than triples in the growth process between bean and sprout. Vitamin C, not even measurable in the cooked bean, now amounts to 7 mg per 100 grams of fresh sprouts. That is why eating bean sprouts is a crunchy way to get your vitamins.

How the nutrient content of foods differs from one part of a plant to another is summarized in the tables that follow. Foods included in these tables were selected at random, the only objective being to represent a variety of edible seeds, leaves and roots.

In Table I you will see that apples contain almost no fat (0.1 gram (.0035 oz) in a 100 gram (3.5 oz) portion), oatmeal contains about 9 per cent fat and nuts (Brazils and walnuts) contain about 68 per cent fat. By examining figures for the types of

Table II: Mineral content in selected foods (100 gram (3.5 oz) portions)

Food	Na mg	K mg	Ca mg	Mg mg	P mg
Apple, raw	1	63	3	2	5
Banana	1	400	6	34	28
Figs, dried	62	970	250	80	89
Spinach, boiled	120	230	160	34	28
Quorn*	240	N	N	N	N
Potatoes, baked (whole)	12	630	11	32	68
Onions, raw	3	160	25	4	30
Brazil nuts, unroasted	3	660	170	410	590
Walnuts	7	450	94	160	380
Brown rice, boiled	3	250	10	110	310
Oatmeal, quick cook, raw	9	350	52	110	380
Olive oil	tr	N	tr	tr	tr
Parsley, fresh	33	760	200	23	64

   * Quorn is a high protein food produced from fungus and used as a meat
    substitute
  ** This figure can vary from 230–5300 mcg per 100g, depending on the
    source of the nuts
***figures enclosed in (-) are estimated values

fat present (saturated, unsaturated and polyunsaturated), it is obvious that walnuts are an excellent source of polyunsaturated fats (which include the essential fatty acids); better than either Brazil nuts or olive oil, which are rich sources of monounsaturated fat.

Brazil nuts, oatmeal and Quorn – a food produced from a form of fungus – contain valuable amounts of protein, and bananas, figs and rice are good sources of carbohydrate. Many of the foods listed have a high water content. Food is often overlooked as an excellent source of pure, safe water.

In Table II, it is obvious that many of the foods selected are rich in the mineral potassium (K), while few contain iodine (I) and selenium (Se). Nuts are good sources of both. Spinach and parsley, both green leafy foods, are excellent sources of most minerals, as are the 'seed' foods of nuts, oatmeal and brown rice.

Fe	Cu	Zn	Cl	Mn	Se	I
mg	mg	mg	mg	mg	mcg	mcg
0.1	0.01	tr	1	tr	tr	tr
0.3	0.1	0.2	79	0.4	(1)***	8
4.2	0.3	0.7	170	0.5	tr	N
1.6	0.01	0.5	56	0.5	(1)***	2
N	N	N	N	N	N	N
0.7	0.14	0.5	120	0.2	2	5
0.3	0.05	0.2	25	0.1	(1)***	3
2.5	1.76	4.2	57	1.2	1530**	20
2.9	1.34	2.7	24	3.4	19	9
1.4	0.85	1.8	230	2.3	(2)***	N
3.8	0.49	3.3	25	3.9	3	N
0.4	tr	tr	tr	tr	tr	N
7.7	0.03	0.7	160	0.2	(1)***	N

From *The Composition* of Food, fifth edition, reproduced with the permission of The Royal Society of Chemistry and the Controller of Her Majesty's Stationary Office.

Table III, which summarizes the vitamin content in the selected foods, describes the nutritional value of parsley. While we do not eat this herb in the same quantities as spinach, the small portions used in many foods can make a considerable contribution to our nutrition. Substantial quantities of biotin are present in the 'seed' foods, and in Quorn. Nuts and olive oil are good sources of vitamin E, present to protect the delicate double bonds in unsaturated oils, and apples and potatoes are both better sources of vitamin C than spinach. Note that none of these foods contain vitamin D; humans get their vitamin D from the action of sunlight on the skin or from fortified foods.

It is important to recall when reading these tables that the figures represent the contents of 100 grams (3.5 oz) of each food, which includes its normal water content.

Table III: Vitamin content in selected foods (100 gram (3.5 oz) portions)

Food	Carotene mcg	Vit D mcg	Vit E mg	Thiamin mg	Ribo-flavir mg
Apple, raw	(17)	0	0.27	0.04	0.02
Banana	21	0	0.27	0.04	0.06
Figs, dried	(64)	0	N	0.08	0.1
Spinach, boiled	3840	0	(0.71)	0.06	0.05
Quorn*	0	0	0	36.6	0.15
Potatoes, baked (whole)	tr	0	0.11	0.37	0.02
Onions, raw	10	0	0.31	0.13	tr
Brazil nuts, unroasted	0	0	7.18	0.67	0.03
Walnuts	0	0	3.83	0.4	0.14
Brown rice, boiled	0	0	0.3	0.14	0.02
Oatmeal, quick cook, raw	0	0	1.5	0.9	0.09
Olive oil	N	0	5.1	0	tr
Parsley, fresh	4040	0	1.7	0.23	0.06

* Quorn is a high protein food produced from fungus and used as a meat substitute
** figures enclosed in (-) are estimated values

Note: There is no vitamin B_{12} in any of these foods with the exception of Quorn. This vitamin must be obtained from animal products, fortified foods, or some form of algae.

Food pyramids

As is obvious from the preceding tables, foods are complex mixtures of different types of nutrients. That makes it difficult to prescribe exactly how you should combine them in a healthy diet. To obtain all the nutrients you need, you must enjoy a diverse combination of foods. An oriental philosophy suggests you eat a little of 30 different foods (ingredients) each day for health and wisdom. Good advice, as long as those foods include fruit, leaves, roots and seeds.

To help us select a healthy balance of foods, nutritionists have spent considerable effort trying to create a schematic representation of an ideal diet. Attitudes towards food change, so

THE PLANTS WE NEED TO EAT

Niacin mg	Vit B6 mg	Vit B12 mcg	Folate mcg	Pantothanate mg	Biotin mcg	Vit C mg
0.1	0.06	0	5	tr	1.2	14
0.7	0.29	0	14	0.36	2.6	11
0.8	0.26	0	9	0.51	N	1
0.9	0.09	0	(90)	0.21	0.1	8
0.3	tr	0.3	7	0.14	9.0	0
1.1	0.54	0	44	0.46	0.5	14
0.7	0.2	0	17	0.11	0.9	5
0.3	0.31	0	21	0.41	11.0	0
1.2	0.67	0	66	1.6	19.0	0
1.3	0	0	10	N	N	0
0.8	0.33	0	60	1.2	21	0
tr	tr	0	tr	tr	tr	0
1.0	0.09	0	170	0.3	0.4	190

From *The Composition* of Food, fifth edition, reproduced with the permission of The Royal Society of Chemistry and the Controller of Her Majesty's Stationary Office.

the ultimate food pyramid has yet to be achieved. None the less, there are some rules for eating that make sense:

* The majority (about 40 per cent of calories) of what we eat each day should come from foods high in complex carbohydrates, such as grain products, including cereals, breads, noodles and pasta. These are also high in complex carbohydrates and B vitamins. Potatoes and other root vegetables also contribute complex carbohydrate to our diet.
* About 30 per cent of our calories should come from fruit and vegetables, although some experts like to separate these two groups. These supply vitamins and minerals that add to those obtained in the previous group.

* Another 20 per cent of calories should be from good protein sources. If these are soya protein, beans, rice and other plant foods, they will also provide complex carbohydrates and fibre.
* Finally, about 10 per cent of your calories should derive from nuts, seeds and oils: good sources of polyunsaturated fats, vitamin E and minerals.

A diet scheme that makes sense and is easy to follow was developed in Australia in 1994.[2] I call it the 'What You Fancy' diet. Each day, eat five portions of starchy food (bread, potatoes, pasta, etc.), four portions of vegetables, three pieces of fruit, two good protein sources, one small helping of seeds or nuts, plus a little of what you fancy.

Other substances in plants

Plants manufacture specific substances to carry out their own life processes, the phytochemicals already mentioned. Many of these do not play a part in the basic life chemistry of humans. Pine trees, for example, produce limonene, a natural insect repellent that fights off the tree-killing bark beetle. These special plant chemicals take many forms, however, and some of them have been shown to have healing properties in humans. For example, substances with molecular structures similar to human oestrogen are believed to ease the symptoms of the menopause and to block the chemical action of substances that cause breast cancer. Still other plant molecules have the ability to cure certain illnesses: derivatives of *Quinghaosu*, an ancient Chinese herbal remedy, are the newest and most promising drugs in the fight against malaria and are currently undergoing medical evaluation for use in Europe and Asia. Although for centuries healers and alternative health experts have respected the curative power of substances in certain plants, intense scrutiny by the scientific community of plant chemicals for curative properties is recent.

Other examples of these miracle molecules are:

* *Brassinosteroids*, plant steroids produced by members of the mustard, or *Brassica* genus, which may provide means of preventing and controlling hormone-sensitive cancers.
* *Carotenoids*, a form of several substances called isoprenoids, which are found mainly in the leaves of plants, where they play an important role in photosynthesis. In humans, many medical experts now believe they may provide powerful protection against free radical damage, and some may help reduce the debilitating physical effects of air pollution, stress and certain viral infections.
* *Bioflavonoids*, sometimes called vitamin P, which were isolated in the mid 1930s from the white part of the skin (pith) of citrus fruit. They are also found in cherries, rose hips, buckwheat, broccoli and apricots, and are thought to aid vitamin C in strengthening capillaries and healing bruising.

WHICH NUTRIENTS ARE NEEDED FOR HUMAN HEALTH?

Before answering that question, consider the fact that it specifies 'human' health. This is because in different animals – from orangutans and opossums, chickens and sheep, to starfish and killer whales – the exact mix of nutrients needed for normal life activities differs from species to species. That is why the science of animal nutrition is so exciting and important.

Scientists have shown that human requirements for specific nutrients also vary according to age, size, sex and the habits and physical condition of an individual. Smokers, for example, require much greater quantities of vitamin C to help block the damage done by free radicals generated by the effects of this habit. People under physical or psychological stress need more vitamin C, B-complex vitamins and essential fatty acids to help maintain their tissues. During menstruation, women require more iron; women planning a pregnancy or in the first trimester of a pregnancy need more folate to support the normal formation of their child's spinal cord and brain. Obviously small children need fewer nutrients than their older siblings. But there are some exceptions: on a proportionate basis, very young, growing children may need more calories from fat than teenagers. Their digestive systems are too small to cope with the amount of carbohydrates needed to meet their energy requirements. As they develop, this changes. In other words, nutritional requirements are not static. However, average figures provide useful guides. The following are suggested requirements for healthy adults enjoying a moderate level of exercise.

There are two types of nutrients: those we need in large quantities, called 'macro-nutrients', and those 'micro-nutrients', which we require in very small amounts. To keep things simple, all essential minerals are included under micro-nutrients despite the fact that some – such as calcium – are required in much larger quantities than others. Some information about the four macro-nutrients – carbohydrate, protein, fats and fibre – is listed below. Detailed information about each of the essential micro-nutrients (vitamins and minerals) follows on page 46, along with Recommended Daily Allowances, often referred to as RDAs.

What are RDAs?

RDAs are derived amounts of specific nutrients experts believe are necessary for good health. They are the scientific basis on which most recommendations about diet and the balance of foods are made. First agreed more than 30 years ago, many now believe they are too low and have considerable limitations because they do not reflect changing needs due to illness, stress and lifestyle variations (the amount of daily exercise is an example).

Scientists arrived at the figures for RDAs in three ways: first, by determining the average quantity of a specific nutrient in the normal diet of healthy people; secondly, by conducting nutritional studies on fit young people to determine the minimum amount of a nutrient required to avoid signs of dietary deficiency; and thirdly, by extrapolation from animal studies (the least accurate form of research).

In other words, the RDA figures – which sometimes are looked upon as absolutes – are more approximation than fact. There are stories about early experts arguing on appropriate numbers. It is most probably apocryphal, but there is a story that two men held different opinions concerning an exact RDA; lunch was approaching, so they compromised by agreeing on an average of the two numbers.

To add to the confusion about how much of each nutrient we need, different expert groups use different measurement criteria and refer to their results by different names. In the United States, RDA stands for Recommended Daily Allowance; in the United Kingdom, the same letters stand for Recommended Daily Amount. There are differences in the numbers assigned to various nutrients under the two schemes. There are also EARs (Estimated Average Requirements) and RNIs (Reference Nutrient Intake – an amount high enough to meet almost everyone's needs and very similar to the old British RDAs). And, most recently, European experts are devising a set of numbers to meet their concept of nutritional requirements.

In this book, the RDAs given are based on the system used in the United States (Recommended Dietary Allowances). Some experts believe these are probably low for normal active people, but they provide a standard, widely accepted benchmark for assessing the foods we should consume.

We assume the average adult woman requires 2,000 calories per day and the average adult males requires 2,500.

MACRO-NUTRIENTS

There are five categories of food substances, or nutrients, we need to consume in large quantities each day. These are water, carbohydrate, protein, fat and fibre.

WATER

Next to oxygen, humans need water to survive. About 70 per cent of an infant's body and 60 per cent of an adult's consists of water. An adult should drink about 3 litres (5.25 pints) of water a day, more if involved in heavy work or sports activity.

Fruit and vegetables – melons and cucumbers for example – are good sources of healthy fluid. Drinks containing caffeine (colas, tea and coffee) act as mild diuretics and can actually increase our daily need for fluid.

CARBOHYDRATES

Carbohydrates, produced during photosynthesis in green plants, are the body's preferred source of energy, although both fats and proteins can also be used. The forms of carbohydrate are:

- *simple sugars*, found in table sugar, honey, molasses and fruit sugar
- *starches*, found in grains, pulses (legumes), fruits and vegetables
- *fibre*, i.e. cellulose and other substances in plants that are not broken down into usable energy during digestion

RDA:
- Adult women, moderately active: 243 grams (8.5 oz)
- Adult men, moderately active: 319 grams (11.3 oz)

Needed for:
- Energy, muscle activity

Good plant sources are:
- Wholemeal bread, legumes, grains, fruit, vegetables (especially root varieties like potatoes and carrots)

PROTEIN

After water, protein is the second most plentiful substance in the human body, where it forms part of all tissue and organ structures, bones, hair and fingernails included. Proteins also form enzymes (metabolic catalysts) and certain hormones (insulin, for example). They consist of very specifically sequenced links of sub-units called amino acids. Twenty amino acids occur in humans, half of which can be synthesized by the body. The others must be obtained on a regular basis from the food we eat. The 'essential amino acids' are: isoleucine, leucine, lysine, methionine, phenylalanine, threonine, tryptophan and valine. Histidine and arginine, sometimes referred to as 'semi-essential', are amino acids that can be synthesized, but at low levels; outside sources are beneficial.

If all of the essential amino acids are not available, the body *limits* the amount of protein made. The total protein you eat is only as useful to your body as the least amount of any one essential amino acid; a deficiency in any one essential amino acid acts as a brake on all protein metabolism. Over time this can result in tissue damage and failure of body processes controlled by enzymes and hormones constructed from protein. That is why vegans must be certain they eat the right blend of foods.

RDA (complete protein):
- Adult women: 37 grams (1.3 oz)
- Adult men: 45 grams (1.6 oz)

Needed for:
- Building muscles; enzymes; repairing and replacing cells

Good plant sources are:
- The best sources of concentrated protein are animal products, including eggs and milk. Although plants also contain proteins, they tend to be deficient in lysine, methionine and tryptophan. By mixing plant sources, however, a good amino acid balance can be achieved: sunflower seeds and rice, soybeans and corn, wheat and rice are healthy combinations. Use the following as a general guide:

Table IV

Plant source	Missing amino acids
Maize (corn)	threonine, tryptophan
Grains	lysine
Beans and pulses (legumes)	methionine, tryptophane
Rice	threonine, trypthohane
Soya beans	methionine (low)

Did you know?
- Cell growth and regeneration depends on an adequate supply of proteins.

- Enzymes control most of the biological activity in the body; some cells contain about 1,000 different enzymes.
- Protein contains nitrogen and in this way differs from other macro-nutrients.

FATS

Oils, butter, beeswax and cholesterol are all forms of fat and share two characteristics: they are insoluble in water and soluble in fat solvents, like benzine and toluene. The fats in foods are almost all in the form of 'triglycerides' and are concentrated sources of calories, containing about twice the food energy as equal weight of carbohydrate or protein.

Although fats have been given a bad name, they are a necessary part of our daily diet.

World Health Organization recommendations:
- Adults: No more than 10 per cent of total daily caloric intake should be derived from saturated fats; the rest should be in the form of mono- or polyunsaturated fatty acids *(see pp.168 and 213–14 for definitions)*. A good goal for good health is to consume no more than 30 per cent of your total caloric intake in the form of fats. Women proportionately need slightly more fat in their diet than men.

  Women – between 20 and 30 per cent total calories as fat (approximately 75 grams (2.6 oz))

  Men – between 15 and 30 per cent total calories as fat (approximately 99 grams (3.5 oz))

Needed for:
- Stored energy, cell membrane structure, absorbing fat-soluble vitamins from the gut, manufacture of cholesterol and steroid hormones, transport of cholesterol, conservation of body heat, protection of kidneys and other organs in the body.

Good plants sources are:
- Seeds, nuts, oils from seeds and nuts, green leaves and the ancient food Spirulina, or blue-green algae

Did you know?

- Saturated fats are dead calories: although they are an excellent source of concentrated energy and form certain structures within the body, they do not have the physical structure (known as 'double-bonds') needed to perform in many of the body's complex biological processes.
- Certain polyunsaturated fats *(see pages 167–8 and 213)* help build our nervous system, help clear arterial walls and carry off cholesterol for proper use and disposal, and form parts of the chemical messenger systems that maintain the communications between the cells in our bodies. No saturated fat can ever perform these activities.
- Olive oil is high in monounsaturated fats, which may help control heart disease. Walnuts are rich in both the omega-3 and omega-6 forms of polyunsaturated fats needed for good health.

Make sure your diet contains a good supply of vitamins E and C to protect the valuable unsaturated fats in your diet. These vitamins are natural antioxidants that keep free radicals from destroying the biologically active parts of these fats. As you increase the amount of polyunsaturated fat in your diet, make sure you also increase your intake of antioxidants.

FIBRE

There are two forms of fibre: soluble and insoluble. Both consist of complex carbohydrates that form structural parts of plants, but which human digestion cannot break down into sugars for use as energy. Therefore they pass from the body without adding calories. In cows and other animals with a complex, two-stomach digestive system (ruminants), grasses and other plants containing large quantities of cellulose (a major form of fibre) can be digested and provide energy.

RDAs:
- Adult women and men: 18 grams (0.6 oz)

Needed for:
- Normal bowel function

Good plant sources:
- Vegetables, fruit, wholegrains and seeds

Did you know?
- Soluble fibre may help lower levels of cholesterol in the blood. Good sources are oats and barley.
- Too much fibre can speed the passage of food to the point where all of its nutrients cannot be removed in the gut.
- A high-fibre diet has been shown to reduce the risk of colon cancer because it speeds the passage of potentially harmful substances through the gut. We need to get the balance right.

MICRO-NUTRIENTS

The following summaries of information on individual micro-nutrients include RDAs *(see p.40)*, a list of primary biological functions and a list of plants that are good sources of the nutrient. Entries for the vitamins also list some of the conditions (such as smoking cigarettes) that increase the need for the nutrient. General summaries about the ways vitamins and minerals work in the body precede the entries for each of these groups of nutrients.

THE VITAMINS

The human body manufactures most – but not all – of the substances needed for energy, digestion, growth, repair and reproduction. In addition to the essential amino acids and essential fatty acids mentioned above, there are other essential organic molecules that cannot be produced in the body and must be obtained, in minute quantities, from food: these are called 'vitamins'. They complete the mix of biological ingredients the body needs to accomplish very specific tasks – such as splitting a hydrogen atom off a molecule or linking up a series of amino acids to form the protein insulin.

When a vitamin is missing from the diet, a series of physical and, in the case of certain vitamins, mental symptoms develop. Unless these have reached a critical stage, most can be reversed by adding the missing nutrient to food. Even then, proper therapeutic vitamin therapy can reverse most changes. Because most vitamins take part in biological activity that affects more than one kind of cell or body tissue, deficiency symptoms may not occur in every person in the same way. This can make diagnosis of an illness difficult, because patients present themselves to their doctor with conditions that appear to have very different origins.

For example, beri-beri is a life-threatening illness caused by a deficiency of vitamin B_1, or thiamin, required for the metabolic breakdown of carbohydrate. Two forms of beri-beri are known: 'wet' beri-beri, in which water retention in tissues is so gross the patient appears to be obese, and 'dry' beri-beri, which affects the nervous system and can produce mental symptoms that suggest madness. Both of these conditions will show improvement within hours of treatment with thiamin. Two patients with what appear to be very different illnesses can be treated quickly and effectively with the administration of a simple nutrient, something we take for granted in the food we eat. Most beri-beri is seen in populations limited to a diet dominated by polished rice. Supplementing the rice, or eating brown rice instead, eliminates the illness.

The link between substances in food and the prevention and cure of specific illnesses has been known for many thousands of years. The earliest medical records, the Ebers papyri, were written before 1500 BC and they clearly define the link between eating liver – which is where vitamin A is stored – and the treatment of night blindness. Galen (AD 130–201), medical advisor to Roman gladiators, spent considerable time studying which foods improved athletic ability. For many centuries military commanders knew there was a link between the rations available for their men and the presence or absence of disease, but there was little formal effort to identify the cause. In many instances the cause was a deficiency of vitamin C, which results in a debilitating and potentially lethal illness: scurvy. Usually outbreaks occurred during long marches or sea voyages when food rations were limited to some form of biscuits or other dried foodstuff. Officers noted that men could be at the point of death, be put down on land or given fresh food, and regain health within days. But where, and what, was the magic food?

Finally, in 1753 John Lind published a study of scurvy, which was one of the earliest documented controlled medical experiments. During a voyage of the British ship *Salisbury*, groups of men suffering from scurvy were each given different treatments: sea water, vinegar, dilute sulphuric acid, 'cyder', and a ration of oranges and lemons. Those given citrus fruits recovered in days and as a consequence, some years later, the British Navy regularly supplemented the diets of their sailors with limes: thus the nickname 'limey'. (Limes contain less vitamin C than lemons but were cheaper at that time.) Why eating citrus fruit prevents scurvy was not known until much later. In the late 1920s, Albert Szent-Gyorgyi isolated and published a paper describing vitamin C (asorbic acid), a substance we now know not only prevents the many symptoms of scurvy, but is a crucial part of the body's antioxidant defence against damage by excess free radicals. In 1937 he was awarded a Nobel Prize for this important work.

A story of serendipity surrounds Szent-Gyorgyi's discovery. During his early work on vitamin C he could not get enough crystallized vitamin for his work. It is claimed that one night his wife was far too heavy handed with the paprika (Hungarian red pepper), and as a consequence the dish prepared for their evening meal was inedible. But something about the taste of the food interested Szent-Gyorgyi and he took his plate to his laboratory to investigate its contents. He found it was rich in vitamin C! He then crystallized the nutrient and proceeded with his experiments with a good supply on hand in an ordinary spice bottle. The message in this story? Never take the nutrient value of any food for granted; the spices and herbs we use add more than flavour and aroma to our food.

Once the critical nature of vitamins and other essential food substances was understood, there was a great rush of research to identify these 'magic molecules' and define their role in human health. By the middle of the twentieth century, around the time of the Second World War, many scientists thought most of the work was done.

There is no doubt that the massive excitement and enthusiasm surrounding the discovery of vitamins and their metabolic importance was crucial to the control and elimination of much human suffering. But it may have reduced our respect for the food in our shopping baskets. Fruit, vegetables and wholegrain bread were now described more as convenient containers of magic molecules than as sweet smelling and delicious sources of pleasure. Perhaps this is where things went wrong. Perhaps when we began disassociating natural pleasures from the food we bring to the table we left the door open to those who wanted to introduce newer, faster, more convenient ways to feed us and our families. Slowly our appetites were subverted, and came to depend on processed foods which had been stripped of their goodness and character. Instead of a diet dominated by fresh wholefoods, we came to eat fatty stodge.

None the less, to understand nutrition and deal with the health and medical problems we now suffer, we must understand both the vitamins and minerals required for the normal biological processes of the body.

There are two types of vitamins: those that dissolve in water and those soluble in fat. Fat-soluble vitamins are stored in body tissues, making it possible to go for some time without 'topping up'. (We need fat in our food to aid absorption of these vital micro-nutrients.) Water-soluble vitamins not used soon after they are absorbed into the body are washed away in the urine or faeces. For this reason, we need to restock our supply of water-soluble vitamins each day.

Note: Suggested RDAs for various age groups are presented in the pages that follow. Lower numbers in a set are appropriate for younger members of an age group.

Fat-Soluble Vitamins

Vitamin A (Retinol)
RDA:
- Requirements are dependent on body weight. The RDA for vitamin A is expressed in 'retinol equivalents' (RE), where 1

RE equals 1 mcg of retinol or 6 mcg of beta-carotene. As the vitamin is stored in the liver, daily doses are not necessary. For adults, an average daily intake of between 500 and 600 mcg of vitamin A, or 1,000 to 1,200 mcg of beta-carotene, will maintain healthy levels in the body. Prolonged and excessive doses of vitamin A may be toxic.

Needed for:
- Healthy skin and hair; normal skin growth and repair
- Normal night vision
- Building cells needed for a strong immune system
- Repair of wear and tear on internal organs, particularly the digestive, respiratory and urinary systems
- Strong teeth and bones

Symptoms suggesting deficiency:
- Acne and a skin condition known as follicular keratinosis
- Night blindness and inflammation of the eyes
- Cystitis, thrush and other infections
- Dandruff
- Mouth ulcers
- Impaired growth
- Reduced levels of sex hormones
- Poor tooth formation – crooked teeth

Conditions contributing to deficiency:
- Smoking
- Excessive alcohol
- Excessive coffee and other drinks containing caffeine

Good plant sources:
- Brightly coloured fruit and vegetables
- Vegetables: Broccoli, cabbage, corn, peas, watercress, carrots, spinach, kale, asparagus, alfalfa sprouts, turnip tops (greens), beetroot tops, mustard tops, pumpkins, dandelions and tomatoes
- Fruit: Orange varieties, including apricots, papayas, peaches, melons

Did you know?:
- Vitamin A is stored in the liver; massive doses may be toxic and should be avoided.

THE PLANTS WE NEED TO EAT

- This vitamin fights infections in the urinary and respiratory tract.

Deficiency may encourage the formation of kidney and gall-stones, and may contribute to problems of the respiratory system such as catarrh, coughs and colds, and sinus trouble.

Beta-carotene

The substance (a provitamin) the body uses to manufacture vitamin A, beta-carotene, is found only in plants. It is rarely toxic. It is a powerful antioxidant – foods rich in beat-carotene are believed to reduce the rates of diseases caused by free radical damage, including certain forms of cancer, heart disease and cataracts. However, this provitamin seems to work best when consumed as part of natural food, rather than in a purified form such as a food supplement. Several intervention studies, in which very large groups of people ate diets with beta-carotene supplement, did not show the reduced rates of disease expected by researchers. This seems to underline the importance of ingesting vitamins as part of wholefoods.

Research suggests that foods containing beta-carotene may be a powerful weapon against cancer. Colon and rectal, bladder, breast, lung and oral cancers have all been shown to have a link with low levels of beta-carotene in the diet.

Good plant sources are:
- Red, orange and green fruits and vegetables: spinach, red and green peppers, pink melons and parsley are examples; the blue-green algae, Spirulina

Vitamin D

Ten different forms have been identified, including Ergocalciferol, derived from plants.

RDA:
- Infants to 1 year: 7.5 to 10 mcg (300–400 IU)
- Children to 10 years: 10 mcg (400 IU)
- Children and adults to 24: 10 mcg (400 IU)
- Adults 25 and older: 5 mcg (200 IU)
- Pregnant women: 10 mcg (400 IU)
- Lactating women: 10 mcg (400 IU)

In the presence of sunlight, vitamin D is produced in the human skin through the conversion of a provitamin.

Needed for:
- Absorption of calcium and phosphorus
- Bone development and repair
- Normal levels of calcium in blood

Symptoms suggesting deficiency:
- Rickets in children and osteomalacia in adults (due to lack of bone rigidity)
- Bowed legs, knock knees, beadlike swellings on ribs
- Backache (sign of some bone diseases)
- Hair loss
- Joint pain and stiffness
- Tooth decay
- Muscle cramps

Counter conditions:
- Saturated fats and fried foods
- Limited exposure to sunlight

Good plant sources:
- Oily fish, fortified milk and other animal products are the richest sources of vitamin D; however, dark green leafy vegetables and some mushrooms contain tiny amounts. Exposure to sunlight causes precursors in the skin to convert to vitamin D.

Did you know?
- Vitamin D is best known for its effect on bone. Rickets (a crippling disease in which calcium fails to deposit in bone and the weight-bearing bones in the legs become bent) was once common among children. When it was discovered that vitamin D deficiency caused the problem, food, specifically milk, was fortified and rickets was conquered. Osteoporosis, like rickets, is characterized by abnormally low levels of calcium in bone tissue. Studies have shown improvement in bone density when calcium supplements are given with vitamin D supplements.

- Most of the vitamin D we need is produced by the action of sunlight on our skin.
- Vitamin D may also help reduce the risk of colon cancer, although the exact mechanism for its action is unknown.

Vitamin E (Tocopherol)
RDA:
- Infants: 3–4 mg
- Children: 6–7 mg
- Adult men: 10 mg
- Adult women: 8 mg (increased by 2 mg during pregnancy and 4 mg when breast feeding)

Needed for:
- Antioxidant activity – vitamin E helps protect cells and their internal structures from attack by free radicals *(see pages 30 and 45)*. Specifically, along with some body enzymes and in conjunction with vitamin C, it helps protect essential fatty acids from oxidation.
- Protection and stabilization of cell membranes
- Protection against heart disease, especially for women
- Protection of the lining of the lungs against damage from air pollution
- Increasing the body's capacity to use oxygen
- Preventing abnormal blood clots
- Reducing the risk of certain forms of cancer
- Protecting the body's store of vitamin A and iron
- Fighting ageing and aiding wound healing
- A healthy reproductive system

Signs of deficiency include (vitamin E deficiency is difficult to diagnose):
- Slow wound healing
- Susceptibility to bruising
- Feeling tired
- Low muscle tone
- Drop in sex drive

Contrary factors:
- Birth control pills

- Saturated fats and fried foods
- Air pollution
- Smoking

Works best in combination with:
- Vitamin C
- Selenium

Good plant sources:
- Safflower oil, nuts, wholegrain cereals, oils from nuts and seeds, wheatgerm oil, sunflower seeds, sweet potatoes

Did you know?
- Vitamin E is removed from grain during the milling process.
- Vitamin E is removed from oils during rigorous refining processes; the resulting residue may be so rich in this vitamin it can be sold on to produce vitamin E supplements.
- Vitamin E and linoleic acid (an essential fatty acid) are found together in Nature, so if you enjoy foods high in linoleic acid – safflower oil, for example – they should also contain vitamin E.

For years scientists could not agree on whether or not vitamin E played a role in maintaining human health. Originally identified during the 1920s as a substance needed for normal reproduction in rats, it was first thought that it had a similar value in humans. As no experimental data conclusively showed this to be the case, many believed vitamin E provided no nutritional benefit.

Vitamin F
This is an old-fashioned term sometimes used to describe essential fatty acids. It is used occasionally in descriptions of the contents of cosmetics. *For information about essential fatty acids, see pages 165–8 and 213.*

Vitamin K (Phylloquinine, Menaquinone)
RDA:
- Infants: 5–10 mcg
- Children:
  1 to 6: 15–20 mcg

7 to 10: 30 mcg
11 to 15: 45 mcg
15 to 18 (boys): 65 mcg
15 to 18 (girls): 55 mcg
- Men 19 to 24: 70 mcg
- Women 19 to 24: 60 mcg
- Men 25 and older: 80 mcg
- Women 25 and older: 79 mcg

Needed for:
- Blood clotting

Signs of deficiency:
- Abnormal bleeding times in the newborn

Good plant sources:
- Green vegetables: alfalfa, asparagus, French beans, broccoli, Brussels sprouts, cabbage, cauliflower, watercress

Contrary factors:
- Antibiotics
- Laxative abuse or excessive use of colonic irrigation
- Problems involving the normal flora (bacterial content) and condition of the large gut

Did you know?
- Humans do not synthesize this vitamin, but can absorb adequate quantities from a combination of the vitamin K synthesized by bacteria in the large intestine and that absorbed from food.

Water-Soluble Vitamins

B-complex vitamins

This group of vitamins includes vitamin B_1 (thiamin), vitamin B_2 complex (riboflavin, nicotinic acid, pyridoxine, pantothenic acid, biotin, inositol, folic acid (folate)) and vitamin B_{12} (cyanocobalamin). None are stored in the human body and any excess is excreted in the urine. All B vitamins are important for the normal growth and maintenance of the central nervous system, are found in the same foods and are easily destroyed by overcooking in too much water. To get the most vitamin B complex from the foods you eat, enjoy them raw or lightly stir-fried.

Symptoms of minor but chronic vitamin B complex deficiency are varied and include fatigue, dermatitis and complications involving the digestive tract. More extreme deficiency results in fever, susceptibility to infections, serious intestinal symptoms such as diarrhoea and signs of central nervous system deterioration such as memory failure, nervousness, poor concentration and delirium. Initial symptoms vary depending on the general health and nutritional status of an individual.

Good plant sources:
- Spirulina, other blue-green algae and yeast extracts. To a lesser degree: wheat, brown rice, soya beans, millet, rye, oatmeal, sunflower seeds, sesame seeds, green leafy vegetables, beans, peas, almonds, peanuts, lentils and kelp. Beers contain some B vitamins, but the amounts vary between types and brands.
- Vitamin B_{12} is most highly concentrated in meat; therefore, to protect their health, vegans and vegetarians should make sure their diets contain adequate supplies from other sources.

Vitamin B_1 (Thiamin)
RDA:
- Infants to 1 year: 0.4 mg
- Children:
  1 to 3: 0.7 mg
  4 to 10: 0.9–1.0 mg
- Males:
  1.2–1.3 mg except during the reproductive years (15 to 50) when 1.5 mg is the suggested requirement
- Females:
  1.0–1.1 mg, except during pregnancy and breast feeding, when 1.5–1.6 mg is the suggested requirement

Needed for:
- Energy production
- Function and repair of the central nervous system
- Conversion of carbohydrate into fat

Symptoms suggesting deficiency:
- Beri-beri *(severe deficiency, see pages 47)*
- Mild neurosis
- Poor concentration
- 'Busy' or 'tingly' legs
- Tingling hands
- Rapid heartbeat
- Tender muscles
- Stomach pain
- Constipation
- Poor memory
- Irritability

Conditions contributing to deficiency:
- Antibiotics
- Birth-control pills
- Food processing and cooking
- Stress
- Caffeine
- Alcohol abuse
- Some antibiotics
- Alkaline agents in food (e.g. baking powder)
- Excess fat in the diet
- Prolonged fasting or starvation

Works best with:
- Other B vitamins
- Manganese
- Magnesium

Good plant sources:
- Whole wheat and enriched cereals, nuts, peas, beans, dried fruit, avocados, cauliflower, spinach

Vitamin B$_2$ (Riboflavin)
RDA:
- Infants: O.4–0.5 mg
- Children:
  1 to 6: 0.8–1.0 mg
  7 to 10: 1.2 mg

- Males:
  11 to 14: 1.5 mg
  15 to 50: 1.7–1.8 mg
  51 and over: 1.4 mg
- Females:
  11 and older: 1.2 mg (increase by 0.3 mg if pregnant and 0.5 mg if breast feeding)

Needed for:
- Energy release from food
- Synthesis of amino acids and fatty acids
- Healthy mucous membranes in the mouth, nose and throat
- Production of hormone by the adrenal glands

Moderately good plant sources:
- Dark green leafy vegetables, mushrooms, asparagus, Brussels sprouts and avocados. Vegans and vegetarians should ensure their diet includes brewer's yeast, fermented yeast products, wholegrains, wheatgerm and enriched cereals. (Milk, eggs and liver are the best sources of this vitamin.)

Signs of possible deficiency:
- Skin eruptions, dermatitis, eczema
- Cracks at the sides of the mouth
- Split and damaged nails
- Sensitivity to light and burning eyes
- Sores on the tongue
- Depression, hysteria
- Retarded growth
- Malformation in infants

Works best with:
- Other B vitamins
- Selenium

Conditions contributing to deficiency:
- Excessive alcohol consumption
- Contraceptive pill
- Caffeine
- Some food preservatives

- Alkaline agents in food (e.g. baking powder)
- Processing foods (see below)

Did you know?
- Exposure to light and heat destroys vitamin B_2; this is why fresh vegetables should be stored in dark, cool places. Riboflavin leaches out into cooking water, so use as little as possible.
- No specific deficiency disease has been associated with a deficiency of vitamin B_2 because it acts with other B vitamins.
- Unlike some other B vitamins, riboflavin is heat-stable. However, it can be destroyed by alkaline substances.

Pantothenic Acid (Vitamin B_5)
- There are no established requirements in humans, but acceptable levels are:

  Infants and children to 3 years: 2–3 mg
  Children to the age of 10: 3–5 mg
  People older than 10: 5–8 mg

Needed for:
- Healthy nervous system
- Energy production
- Fat metabolism
- Controlling stress
- Healthy skin and hair

Symptoms suggesting deficiency:
- Poor concentration, apathy
- Lack of energy, feeling of exhaustion
- Muscle cramps or 'twitching'
- Burning and tingling feet and hands
- Reduced production of bile
- Tender heels
- Feeling of nausea
- Grinding teeth during sleep

Good plant sources:
- Wholegrain cereals, dried beans, nuts, avocados, sweet potatoes (yellow yams)

Did you know?

- Caffeine, alcohol and stress increase the requirement for this vitamin.
- It needs other B-complex vitamins for absorption.
- Milling and bleaching grains to make flour destroys pantothenic acid.

Vitamin B$_6$ (Pyridoxine)
RDA:

- Infants to 1 year: 0.3–0.6 mg
- Children:
  1 to 6 years: 1.1 mg
  7 to 10 years: 1.4 mg
- Males (11 and older): between 1.7 and 2.0 mg, depending on weight
- Females (11 and older): 1.5 mg
  Pregnant women and nursing mothers should increase this amount by about one half (2.2 mg)

Needed for:

- Protein metabolism
- A healthy central nervous system
- Healthy digestive system
- Helps maintain healthy skin
- Helps convert tryptophan into niacin
- Helps balance sex hormones; thus important for normal menstrual cycle
- Helps in the formation of antibodies and strengthens the immune system
- Helps the formation of red blood cells
- Helps the release of energy from food

Signs of deficiency:

- Depression, irritability and nervousness
- Muscle cramps or tremor
- Dry flaky skin
- Burning or tingling hands
- Water retention
- General lack of energy

Action blocked by:
- Alcohol
- Smoking
- Contraceptive pills
- High protein intake
- Certain prescribed medications
- Certain genetic diseases

Did you know?
- Vitamin B_6 in foods has a calming effect on the nervous system.
- To be effective, it requires the presence of other B-complex vitamins, zinc and magnesium.
- Poor storage and overcooking destroy vitamin B_6 in food.

Good plant sources include:
- Soya beans, walnuts, peanuts (groundnuts), wheatgerm, bananas, Brussels sprouts, dried beans, wholegrains, seeds and most green vegetables

Vitamin B_{12} (Cobalamin)
RDA
- Infants (to 1 year): 0.3–0.5 mcg
- Children:
  up to 6 years: 0.6–1.0 mcg
  up to 10 years: 1.4 mcg
- Adults: 2.0 mcg
  Pregnant women: 2.2 mcg
  While breast feeding: 2.6 mcg

Needed for:
- Protein metabolism
- Oxygen transport
- Energy
- Synthesis of genetic material (DNA and RNA)
- Healthy nervous system

Signs of possible deficiency:
- Anxiety, irritability and feelings of tension
- Constipation or diarrhoea
- Sensitive gums and mouth

- Tender or sore muscles
- Dermatitis or eczema
- Poor hair quality
- Tiredness, dizziness, moodiness
- Loss of appetite
- Numbness, tingling or other sensory changes in hands and feet

Did you know?
- Alcohol, smoking and drugs that block the formation of stomach acid increase the need for vitamin B_{12}.
- This vitamin is not produced by plants, but by bacteria in the gut and micro-organisms used to ferment foods.
- Vegetarians, especially vegans, should make certain their diet includes adequate quantities of this vitamin.

Satisfactory plant sources:
- Fermented soya bean products, fortified yeast extract and fortified cereals

Folate (Folic Acid)
RDA:
- Infants (up to 1 year): 25–35 mcg
- Children:
  1 to 6 years: 50–75 mcg
  7 to 10 years: 100 mcg
- Males:
  10 years to puberty: 150 mcg
- Adults: 200 mcg
- Females: Same as males, but increased to 400 mcg during pregnancy
  While breast feeding, increase to 280 mcg during the first six months and 260 mcg after that

Needed for:
- Early development of the nervous system (to prevent spina bifida)
- General health of brain and nervous system
- Release of energy from food
- A healthy inflammation response

Signs of possible deficiency:
- Poor memory
- Depression, feelings of anxiety and tension
- Poor appetite
- Lack of energy
- Cracked lips
- Poor skin quality, eczema
- Prematurely grey hair

Countered by:
- Contraceptive pill
- Processes used in manufacturing food

Good plant sources are:
- Green leafy vegetables, asparagus, avocado, broccoli, cauliflower, sweet potatoes and darkly coloured yams, cereals, nuts, wheatgerm, kidney beans. Also found in sprouted seeds.

Did you know?
- The name is derived from the word 'foliage'.
- Folate needs other B vitamins (especially B_{12}) to function well.
- Taking excessively large doses of folate can mask a vitamin B_{12} deficiency.

Folate is important for normal development of the human spinal cord. The brain and spinal cord develop from an embryonic structure called the 'neural tube', which fails to close when the mother is folate deficient. Because this happens during the first few weeks of pregnancy, pregnant women should ask their doctors for advice about supplements during this time. Women planning a pregnancy should choose foods rich in this vitamin.

Niacin (also called Vitamin B_3)
RDA:
- Infants to 1 year: 5–6 mg
- Children:
  1 to 3 years: 9 mg
  4 to 10 years: 12.5 mg

- Males:
  11 to 14 years: 17 mg
  15 and older: Up to 19 mg; decrease after middle age
- Females
  11 years and older: 15 mg; decrease slowly after middle age
  increase by 2 or 3 mg during pregnancy and lactation

Needed for:
- Release of energy from carbohydrates and oxidation of fats
- A healthy digestive and nervous system
- Production of sex hormones
- Normal function of the brain, nervous system and digestive system

Symptoms suggesting deficiency:
- Pellagra – a potential fatal disease affecting all parts of the body. Primary symptoms are dry, darkly mottled skin, diarrhoea and mental confusion.

Did you know?
- Niacin deficiency usually occurs in conjunction with deficiencies in the other B-complex vitamins, iron and protein.
- Niacin can be produced by the human body from tryptophan, an amino acid.
- Processing grain to make white flour removes most of the niacin.

Good plant sources:
- Nuts (especially peanuts), wholemeal bread, dried beans

Biotin
RDA:
- Infants to 1 year: 10–15 mcg
- Children:
  1 to 6 years: 20–25 mcg
  7 to 10 years: 30 mcg
- Adults and children older than 11 years: 30–100 mcg

Needed for:
- Creation of essential fatty acids

- Protein metabolism
- Healthy hair and skin
- Folate activity

Symptoms suggesting deficiency:
- Poor condition of hair and skin, dermatitis
- Sore muscles
- Poor appetite

Did you know?
- Biotin is synthesized by microbes in the intestine.
- Raw egg white blocks absorption of biotin from food in the gut.
- Biotin works best with good supply of B vitamins, magnesium and manganese.

Good plant sources:
- Wheatgerm and wheat bran, Quorn, raw onions, leeks, cauliflower, mange-tout, mung beans, blackeyed beans

Vitamin C
RDA:
- Infants and children to the age of 10: 30–45 mg
- Children to the age of 14: 45–50 mg
- Older children and adults: 50–60 mg
- Pregnant women should increase their intake by 10 mg
- While breast feeding, increase by 30 to 35 mg per day
- *Smokers require about twice as much vitamin C as non-smokers*

Note: Many people feel these levels are low, particularly in light of what we are learning about the importance of vitamin C as an antioxidant and its role in the prevention of certain cancers and heart disease.

Needed for:
- Healthy skin, gums and cardiovascular system
- Absorption of iron from food
- Blocking damage from excessive levels of free radicals
- A healthy immune system
- Wound healing and healthy skin

This is a powerful antioxidant that helps protects cell damage by free radicals. It may help prevent cancer, heart disease and other degenerative diseases of old age.

Symptoms suggesting deficiency:
- Scurvy
- Frequent colds and minor infections
- Bruising
- Tender gums
- Slow wound healing
- Patchy skin
- Joint pain

Conditions contributing to deficiency:
- Alcohol, smoking, stress, oral contraceptives, certain medications and air pollution, rheumatoid arthritis, burns, surgery and trauma, cancer

Good plant sources:
- Blackcurrants, strawberries, greens (parsley, broccoli, kale, etc.), citrus fruit, spinach, green peppers, watercress and potatoes

Did you know?
- Vitamin C works in partnership with vitamin E, certain bioflavonoids and vitamin B complex.
- It leaches from food into cooking water.
- It can be deactivated by heat.
- Foods rich in vitamin C are acidic and should never be cooked or stored in aluminium pans or containers.

In 1970, Nobel Laureate Dr Linus Pauling put his international reputation on the line by proclaiming the benefits of vitamin C. Taken in large doses, he claimed this vitamin not only prevented and fought off symptoms of the common cold, but aided healing and extended life. Unfortunately, most attention was focused on the common cold and clinical research results were ambiguous at best and contemptuous at worse. The common cold may be 'common', but its cause and means of clinical expression remain a medical mystery. What would have happened, one must question, if attention had been placed on another aspect of the vitamin C claims: rapid healing of wounds.

Several decades after Pauling's proclamation there are reams of research papers demonstrating the power of vitamin C, along with vitamin E and beta-carotene (all antioxidants), in preventing or suppressing certain types of cancer, preventing heart disease and improving the immune system, thus reducing the danger of infection.

There are reasons to be cautious about these findings, however. To date there is no reason to believe vitamin C will cure cancer once it has taken hold, although the vitamin has been shown to suppress the harmful action of chemicals known to induce cancerous growths. The rate of infection among tested populations fell, although there was little change in the length of illness among those who became ill. And finally, the effects of vitamin C on heart disease probably involve the oxidation of one type of cholesterol (LDL, or Low Density Lipid cholesterol) and vitamin E has been shown to be more effective. It must be said, however, that vitamin C 'assists' vitamin E to do its work.

MINERALS

Minerals – the inorganic substances left after all water and organic substances are removed from living tissue – make up about 4 per cent of the weight of living tissue. Although most minerals are located in the bones, or skeletal parts of the body, they also play indispensable roles in cells and tissue fluids as part of key biological processes, such as the transmission of nerve impulses, the formation of blood, oxygen transport and energy production. Some minerals work with enzymes as cofactors, others have specific functions, such as that played by iron as part of the haemoglobin molecule responsible for the transport of oxygen from the lung to body tissues.

The balance of minerals in living tissues is delicate and complex because certain of these inorganic substances work in combination with each other, some compete with one another for absorption from food and others are antagonists within body tissues. For this reason, it is best to fulfil as much of your mineral requirement as possible from food, where minerals are usually found in balanced combinations. Obtaining minerals from supplements, or unusual food combinations, may result in imbalance. Chronic shortage of any essential mineral will result in symptoms of low-grade or acute deficiency.

Like vitamins, minerals are considered to be 'essential' when:

* their absence in a diet causes a specific symptom, or series of symptoms, which are reversed when the mineral is reintroduced into the food;
* scientists have identified a specific biological purpose of their presence in the human body; and/or
* they have been identified as an essential part of some other nutrient (the presence of cobalt in vitamin B_{12} is an example).

Certain minerals are required in relatively large daily quantities for good health; examples of these are calcium, potassium and phosphorous. Others – zinc, copper and iodine, for examples – are needed is minute amounts. The fact that one mineral is needed in very small quantities and another in much larger amounts does not mean one is less important that the other. A deficiency in any essential mineral will have disastrous results.

As a general rule, the amount of trace minerals in the body is kept in delicate balance by a system of absorption from the digestive tract, circulation in the body and utilization. Extra minerals are excreted either by the kidneys or from the bowel. If this system of limitation breaks down, problems can arise. For example, selenium and zinc are essential for good health, but large amounts can be toxic.

A few minerals found in the human body, or included in lists of required nutrients by some sources, have not been identified as 'essential'. None the less, they are found in a balanced diet and there is reason to believe they play a significant biological role in humans; fluoride and vanadium are examples.

RDAs for minerals

Like vitamins, RDAs or Recommended Daily Allowances for minerals differ for various groups of people. Age and size obviously help determine how much is needed, but so do activity levels, growth rate, pregnancy and breast feeding, illness and recovery from injury. Always remember that RDAs are recommendations and individual needs vary.

Foods contain different quantities and blends of minerals. In Table II *(pages 34–5)*, you will notice that 100 grams (3.5 oz) of boiled spinach contains 160 mg (5.6 oz) of calcium, whereas the same weight of onions contains only 25 mg (0.8 oz). Brazil nuts are very rich in selenium, but only a trace of this mineral will be obtained by eating a similar weight of dried figs. There are only trace amounts of minerals in olive oil and none – other than sodium – in Quorn. This is another reason why a diet based on a variety of foods derived from plants is a wise choice.

Calcium
RDA:
- Infants:
  0 to 0.5 years: 400 mg
  0.5 to 1 year: 600 mg
- Children:
  1 to 10 years: 800 mg
  11 to 24 years: 1,200 mg
- Adults: 800 mg
  Pregnant or lactating women: 1,200 mg

Needed for:
- Strong bones and teeth
- Normal blood clotting
- Muscle contraction
- Transmission of nerve impulses

Good plant sources:
- Dark green leafy plants, broccoli, citrus fruits, dried peas and beans

Did you know?
- Hard water can be a source of calcium.
- Calcium works in balance with phosphorus.
- Healthy bones also need vitamin D, protein and other nutrients.
- The average adult body contains 1.2 kg of calcium.

- Metabolism of calcium is controlled by the parathyroid hormone.
- Calcium and magnesium are both needed for muscle contraction.
- The prolonged consumption of too little calcium plus too much phosphorous coupled with insufficient strenuous exercise promotes the development of osteoporosis.
- Bone is a constantly growing and changing tissue and – unlike the teeth – never loses its need for calcium.
- Some experts suggest oxalic acid blocks the absorption of calcium from food.

Chloride
RDA:
- Infants:
  0 to 0.5: 180 mg
  0.5 to 1 year: 300 mg
- Children:
  1 year: 350 mg
  2 to 5 years: 500 mg
  6 to 9 years: 600 mg
  10 to 18 years: 750 mg
- Adults: 750 mg

Needed for:
- Normal electrolyte balance in body fluids

Good plant sources:
- Most fruits and vegetables; table salt

Did you know?
- Most foods contain chloride, making supplementation unnecessary in most cases.
- Gastric and intestinal secretions contain high levels of chloride.
- Salting food adds chloride and sodium to the diet.
- Prolonged vomiting or diarrhoea can deplete the body's chloride levels.

Chromium
RDA:
- Little is known about human requirements for this mineral. The following are safe ranges of chromium which are based on average intake:

  Infants (depending on age): 0.01–0.06 mg
  Children (depending on age): 0.02–0.12 mg
  Children over 10 and adults: 0.05–0.2 mg

Needed for:
- The metabolism of carbohydrate
- Work with insulin to control glucose levels

Good plant sources:
- Yeast products, wholegrains

Did you know?
- Refining grains removes most of their chromium content.
- Scientific evidence suggests a link between diets based on processed foods and refined sugars and an increased risk of diabetes.

Cobalt
RDA:
- No specific RDA has been suggested for cobalt. Some experts suggest 6–8 mcg per day.

Needed for:
- Vitamin B_{12} activity
- Formation of red blood cells
- Enzyme activity

Good plant sources:
- Buckwheat, figs, green leafy vegetables, watercress, cabbage, some beers *(see the section on vitamin B_{12}, page 61)*

Copper
RDA:
- No RDAs are available for copper, but suggested safe levels are:

Infants (depending on age): 0.4–0.7 mg
Children (3 to 10 years): 0.7–2.0 mg
Others: 1.5–3.0 mg

Good plant sources:
- Wholegrains, nuts, peas and dried beans, fresh and dried fruits

Needed for:
- Normal connective tissue and bones
- Normal nerve tissue and conduction of nerve impulses
- Haemoglobin production
- Normal function of several important enzymes
- Production of melanin in the skin
- Production of white blood cells to fight infection
- Help in the production of phospholipids

Did you know?
- Infants fed only cow's milk after three months of age can develop copper deficiency, which can cause anaemia.
- Copper deficiency may result in reduced skin pigmentation.
- Copper can be toxic in high levels and have serious side-effects.

Fluoride
- After the first six months of life, safe levels of fluoride are thought to range between 0.2 and 3.5 mg per day.

Needed for:
- There is no biochemical activity in the human body that is known to require fluoride. However, this mineral has been found to strengthen teeth and bones.

Good plant sources:
- Seaweed, soya bean, tea leaves. The fluoride content of plants varies and depends on the levels in the soil in which they are grown.

Did you know?
- Medical evidence suggests dietary levels of fluoride may help reduce risk from osteoporosis.
- In many places, fluoride is added to the public water supply

to help lower the rate of dental decay in a population. However, because there is evidence there is a risk of cancer from compounds containing fluoride, some medical experts see a danger in this practice and want it stopped.

Iodine
RDA:
- To prevent goitre (an enlargement of the thyroid gland), about 150 mcg of iodine are needed each day.

  Infants (up to 1 year): 40–50 mcg
  Children 1 to 7 years: 70–90 mcg
  Older children and adults: 120–150 mcg
  Pregnant women: 175 mcg
  Lactating women: 200 mcg

Needed for:
- Prevention of goitre.
- Production of thyroid hormone, which helps control the metabolic rate, growth and normal development.

Good plant sources are:
- Kelp and other forms of seaweed, iodised and natural sea salt, green vegetables grown in iodine-rich soil

Did you know?
- Infants born to mothers with iodine deficiency can suffer from cretinism, symptoms of which include poor bone formation, slow mental development and poor muscle function.

Iron
RDA:
- Infants (depending on age): 6–10 mg
- Children (1 to 10 years): 10 mg
- Men: 10–12 mg
- Women: 15 mg
  Pregnant: 30 mg

Needed for:
- Production of important enzymes, especially those which stimulate metabolism

- Part of myoglobin, an important part of muscle tissue
- Forming the basis of haemoglobin, the chemical structure in red blood cells responsible for the transport of oxygen around the body

Good plant sources:
- Dried fruit, beans, nuts, whole cereal grains, green and leafy vegetables

Did you know?
- Vitamin C is necessary for the absorption of iron, so try to combine iron and vitamin C rich foods in the same meal.
- The chemical structures of the haemoglobin in blood and chlorophyll in the green parts of plants are similar, and both contain iron. If you want more iron in your diet, add more green foods to your meals.
- Iron is one of the most widely recognized vital elements in human nutrition.
- If you eat no meat (a rich source of iron), make sure you enjoy kale and other iron-rich foods, together with those high in vitamin C to help your body absorb the iron.
- High fibre content in the diet, large amounts of foods containing zinc and the tannin in coffee and tea decrease iron absorption.

Magnesium
RDA:
- Sources disagree, however, safe levels are:
- Infants (up to 1 year): 50 mg
- Children:
  1 to 3 years: 80 mg
  4 to 6 years: 120 mg
  7 to 10 years: 170 mg
- Males (11 and older): 300–400 mg
- Females (over 11 years): 300–350 mg
  Pregnant women: 350 mg
  Women nursing for more than 6 months: 340 mg

Needed for:
- Strong bones and teeth

- Transmission of nerve impulses
- Muscle contraction
- Protein and lipid synthesis
- Energy production

Good plant sources:
- All green vegetables; the higher the chlorophyll content, the greater the concentration of magnesium. Also dried peas and beans, citrus fruit, grains, nuts and seeds

Did you know?
- Three months or longer on a low-magnesium diet can cause symptoms which include weakness, nausea, lack of coordination, confusion and gastrointestinal disorders. More prolonged deficiency has been linked to anorexia, skin lesions, abnormal muscle movements and hair loss.
- Foods rich in magnesium have a calming effect on the nervous system.
- Excess levels of magnesium can be toxic, causing lethargy and weakness. This is usually not due to diet, however, but to the overuse of antacids and laxatives containing magnesium.

Manganese
RDA:
- No dietary requirements have been established; however, levels between 2 and 5 mg per day have been found to be safe in adults.

Needed for:
- Fatty acid and cholesterol production
- Protein digestion and synthesis
- Blood clotting
- Synthesis of collagen for strong skin and tissues

Good plant sources:
- Wholegrains, tea, dried beans and peas, nuts, green vegetables, dried fruit, vegetable roots and tubers

Molybdenum
RDA:
- No specific requirements have been established for this mineral; however, levels considered to be safe for adults range from 75 to 250 mg per day.

Needed for:
- Part of certain important oxidative enzymes
- Iron metabolism

Good plant sources:
- Depending on soil content, good sources include whole-grains, leafy vegetables and grains. Hard water is another natural source of this mineral.

Did you know?
- No symptoms of molybdenum deficiency have been identified in humans.

Phosphorous
RDA:
- Infants (0 to 1 year): 300–500 mg
- Children (1 to 10 years): 800–1,200 mg
- Above 11 years and adults: 1,200 mg
- After the mid 20s, about 800 mg is needed except in pregnant and lactating women, who continue to require about 1,200 mg per day

Needed for:
- Strong bones and teeth
- Cell and tissue structure
- Structure of DNA and RNA
- Growth and repair of tissues
- Protein synthesis
- Phospholipid synthesis
- Acid-base balance

Good plant sources:
- Wheat bran, wholegrain, nuts, green vegetables, potatoes baked in their skins

Did you know?

- Phosphorous is the second most plentiful mineral in the body.
- Excessive iron consumption can block the normal absorption.

A diet high in soft drinks, pre-packaged and fast foods may result in an excess of phosphorus in the body. As the balance between this mineral and calcium is critical, excess phosphorous can disrupt the way calcium is absorbed and used by the body.

Potassium
RDA:

- The body's need for this mineral varies with body size, and loss of body fluids through increased activity levels, vomiting, diarrhoea, excessive use of laxatives and diuretics and kidney disease. There are no agreed dietary recommendations available for potassium; however, a healthy adult would probably require 2,000 mg.

Needed for:

- Maintaining the chemical (electrolyte) balance within the body's cells
- Muscle contraction
- Carbohydrate metabolism
- Nerve transmission

Good plant sources:

- Bananas, apricots, other dried and fresh fruit, potatoes

Did you know?

- Sterility, muscle weakness, fragile bones, renal disease, dangerous changes in heart rate, damaged nerve transmission and even death can result from prolonged potassium deficiency.

Selenium
RDA:

- Expert opinions vary. As a general rule 0.1 mg/day is considered safe for healthy adults, but published figures range from

40 to 70 mcg for adult males and 45 to 55 mcg for females. Pregnancy and breast feeding increases this amount by 10–20 mcg.

Needed for:
- Part of the body's natural antioxidant system that fights damage by free radicals
- Reducing the risk of heart disease and certain cancers Maintaining the elasticity of the skin and reducing signs of ageing

Good plant sources:
- Wholegrain cereals

Note: The selenium content in plants varies with the levels of this mineral in the soil. In areas where selenium is depleted, little will be found in crops.

Did you know?
- Changes in eating patterns and the commercial flow of food products between various parts of the world affect the amount of selenium available in the diet. For example, at one time most of the flour used to bake bread in Great Britain came from wheat grown in Canada, where the level of selenium in the soil is high. After the implementation of trade agreements between countries in the European Union, more wheat was purchased from European countries, where the levels are much lower. As the soil in most of Great Britain is low in selenium, thus offering little to replace that lost by the change in the bread, some experts are concerned that the general British diet may now be too low in this essential mineral.
- Excessive amounts of selenium can be toxic and may interfere with bone growth; children growing up in selenium-rich regions tend to have higher levels of dental decay. Excessive amounts absorbed as industrial pollutants can cause damage to the heart muscle and liver ailments.

Sodium
RDA:
- No RDAs exist, although levels ranging between 1 and 3 grams per day are considered safe for normal adults.

Needed for:
- Water balance in the body
- Normal heart rhythm
- Transmission of nerve impulses
- Muscle contractions
- Removal of CO_2 from the body
- Absorption and transport of amino acids

Good sources:
- Table salt is the primary dietary source of sodium, although most foods contain sodium in natural balance with other minerals.

Did you know?
- Medical evidence suggests a link between excessive consumption of sodium and increased risk of high blood pressure and oedema

Zinc
RDA:
- Infants (0 to 1 year): 5 mg
- Children (1 to 10 years): 10 mg
- Males (11 years and older): 15 mg
- Females (11 years and older): 12 mg
  Pregnant: 30 mg
  Breast feeding: 15 mg

Needed for:
- A healthy immune system
- Healthy skin and hair
- Sexual development
- Normal enzyme activity
- Growth and development
- Detoxifying substances in the liver
- Energy production
- Insulin activity
- Normal sense of smell

Deficiency
- Zinc deficiency can result in stunted growth and may play a role in eating disorders

Good plant sources:
- Sea kale and sea vegetables, wholegrains, Brazil nuts

Notes
1 Erdman, J. W. and Fordyce, E. J., 'Soy protein and the human diet', *American Journal of Clinical Nutrition* 49 (1989), pp.725–37
2 Prepared by *Australian Women's Weekly*, 1994

A DIRECTORY OF FOODS
FROM PLANTS

What are foods from plants?

There are two groups of foods to consider. First are the obvious parts of plants: leaves (spinach and kale); roots and tubers (potatoes, carrots); stalks (celery, rhubarb); fruits (apples and tomatoes); and grains, nuts and seeds. We also eat flowers: cauliflower, broccoli and artichokes, for example. True blossoms can also add flavour and visual appeal to food. Brightly coloured nasturtium flowers add style and a peppery taste to salads, while blossoms from herbs – borage and marjoram in particular – make a beautiful garnish. Squash flowers are delicious when stuffed with lightly seasoned squash purée and gently sautéed in butter or nut oil.

The second group of foods from plants comprises those that no longer bear any resemblance to a living thing. Bread, pasta, spaghetti, couscous and all forms of foods made from flour or crushed grain are the most numerous of these. The oil you use to dress a salad and the sugar stirred into your coffee originate from plants. Then there are wines from grapes, cider from apples and beer from hops and grain. Even honey is a food from plants, although the industrious bee must intercede between nectar-laden flowers and what you spread on breakfast toast.

As already explained, foods from plants contain the protective fibre, vitamins, minerals, essential fats and miraculous phytochemicals that bring us vitality and reduce the risk of infections and major killer illnesses like cancer and heart disease. Daily, we can supply our bodies the balanced and complete set of nutrients it needs by enjoying a variety of fruit, wholegrain cereal products (bread, pasta and breakfast cereal), red and green vegetables, and carbohydrate-filled roots, nuts, pulses and seeds. Along with a very modest amount of protein from milk or eggs, or from a carefully balanced diet of plant proteins, everything we need for good health and long life is available to us from plants.

This chapter explains all you need to know about the nutrient content of a range of enjoyable plant foods. Together with the shopping and cooking tips in Chapter 5, it shows how healthy eating can be interesting and can actually help stretch the family food budget.

How to use this directory

Obviously not all foods from plants can be listed in a book this small; they number many hundreds. So those included here are ingredients in European and American cooking. A few 'future foods', such as blue-green algae and kombucha, are also listed because of their considerable nutrient and special healing content.

Foods are listed alphabetically by their British name. The American name follows in brackets, e.g. courgettes (zucchini). Many entries are cross-referenced to major categories of foods, such as grains, oils, herbs and nuts and seeds, which provide general information about the food group along with the details of individual items. Cross-referencing will lead you also to other parts of the book for additional facts.

Entries vary in length; this does not suggest a preference for one food over another, only that some foods have particularly interesting histories.

What to look for when shopping

In general *(also see Chapter 5)*:

Squeeze items gently. If potatoes, carrots or turnips feel rubbery, they are old and probably lack taste. Pass them by.

Many types of fruit, particularly the citrus fruits, are sprayed with a fine wax to extend their shelf-life. This is difficult to wash off. If you plan to use fruit rind in a drink or food, ask the store manager for unsprayed varieties.

Avoid leafy vegetables with holes; you will be taking home an uninvited guest.

Green patches on carrots and potatoes are toxic and should be avoided.

Finally, if you find white powder on an item, rub it with your finger. If it does not come off, it is probably a natural part of the plant and should be ignored. (Broccoli and cabbage both have white 'bloom' on their leaves and stalks.) If, however, the white patch rubs off, it may contain residues from an agricultural chemical. This sometimes happens when food is sprayed too close to harvest. Find something else for dinner and speak to your store manager.

❋ A DIRECTORY OF FOODS FROM PLANTS ❋

ALFALFA SPROUTS

Rich source of:
- Most B vitamins, amino acids, especially tryptophan

Also contain:
- Water, some protein, folate, carotene, riboflavin, pantothenate, vitamin K, potassium, magnesium, phosphate

Health benefits:
- High in nutrients and fibre and low in fat and calories.
- May contain substances that help lower blood cholesterol levels.
- Some evidence suggests that vitamin K helps prevent the loss of calcium and reduces the possibility of osteoporosis.

Facts and tips:
- Once thought suitable only as animal feed, alfalfa has found modern popularity as a dietary source of vitamins not often found together in the same foods. You can take advantage of this special nutrient 'packaging' by sprouting the seeds at home and using the young green leaves as added flavour and texture in sandwiches and salads. Or, if you use a juicer, add sprouts when preparing a vegetable blend.

ALGAE (SPIRULINA AND CHLORELLA)

Rich source of:
- B vitamins (including B_{12}), essential fatty acids, including GLA (gamma-linolenic acid), iodine and chlorophyll

Also contain:
- Protein (good source of amino acids), vitamins C, and E, beta-carotene, calcium, iron, potassium, magnesium, phosphorus, zinc

Health benefits:
- Algae have been shown to help fight anaemia.
- They may help boost the immune system and protect against free radical damage because of their antioxidant content.
- A highly concentrated balance of protein and other nutrients in a low-calorie form.

Facts and tips:
- May be an important food in the twenty-first century; contain most required nutrients in a balance that helps us grow and helps us heal.
- Health food outlets are currently the best places to buy Spirulina and Chlorella.
- (*See* Sea Vegetables for information about other plants from the sea.)

ALMONDS
See Nuts

APPLES

Rich source of:
- Pectin (a soluble fibre); carbohydrate for energy and phytochemicals
- Vitamin C, although amounts vary with variety and time at which apples were picked; buy them as close to the tree as possible
- Water

Also contain:
- Potassium, some calcium, magnesium, phosphorous, iron and copper, carotene, vitamin E, thiamin, riboflavin, niacin, vitamin B$_6$, biotin

Health benefits:
- One of Nature's miracle foods. Your grandmother was right: an apple a day can help keep the doctor away.
- Apples help prevent both constipation (raw) and diarrhoea (stewed). Relatively low in calories, but rich in fructose (a type of sugar that is broken down slowly by the body and helps maintain blood sugar levels), apples help smooth out blood sugar levels, thus avoiding the roller-coaster peaks and dives that can result in feelings of weakness, hunger, dizziness and slight confusion. (These are the symptoms of hypoglycaemia. If you experience them often, see your doctor.)

  Eating two or more apples a day is thought to relieve arthritis, tension headaches, stress and asthma.

Facts and tips:
- Apples are a well travelled fruit. First brought to Great Britain and other European countries by the Roman conquerors before the birth of Christ, they eventually found their way to the New World in 1628, when the Englishman John Endicott first imported apple seeds for planting. The food and medicinal value of apples, both fresh and dried, was soon recognized and as people moved west they took a bag of apple seeds with them. The American legendary figure Johnnie Appleseed (Jonathan Chapman) and others like him spread seeds to plant a million apple trees across the face of America. Soon, to be truly American was to be 'as American as apple-pie'.

  Apples vary widely in texture and sweetness. Try several varieties and see which your family enjoy raw and cooked. Get to know the colour, smell, texture and taste of apples.

  Pre-packaged apples may save a few pennies, but often include bruised and misshapen fruit. Select your own. It takes time, but is worth the effort. Chose firm apples free of soft spots or browning around the central core.

  Dried apples, a favourite since the Stone Age, are a healthy treat that takes up very little space and comes packed with energy, fibre and iron.

APRICOTS

Rich source of:
- Fibre, carotene, carbohydrate energy (dried fruit has very high potassium content)

Good source of:
- Vitamin C (best in fresh fruit), iron

Warning:
- Some dried apricots contain sulphur dioxide (E220), a preservative, which can cause asthma attacks in susceptible people; read the package list of ingredients carefully before you buy.

Health benefits:
- Apricots have good antioxidant properties to fight degenerative illnesses.
- Major research studies show lower cancer levels among populations eating large quantities of foods rich in beta-carotene.
- Fibre helps prevent constipation.

Facts and tips:
- Sometimes recommended to pregnant women and nursing mothers because of the combined effects of energy, vitamins, potassium and fibre.

ARTICHOKES

A delicious and much misunderstood food!

Good source of:
- Fibre, vitamin C, potassium and folic acid

Also contain:
- Niacin and other B vitamins; some vitamin E and C

Health benefits:

- This popular food in France and Italy is also rich in phyto-chemicals believed to be good for the liver, to help control gallstones and to control cholesterol levels. As the liver is the primary source of cholesterol circulating in the blood, this makes sense. The substance thought to control cholesterol is cynarin, found in the base of the leaves. Pharmacies in France sell capsules of artichoke leaves to be taken to help liver ailments.

Facts and tips:

- True artichokes are the flower buds of a type of thistle and grow in rocky, windy places where little else will form a cash crop. Driving through Brittany you will find fields full of ungainly looking plants with long stalks reaching upwards; these stalks support the flower buds seeking the sun before opening.

  Artichokes once took pride of place in Renaissance gardens. It is said that Catherine de Medici once ate so many at a wedding feast she nearly burst, and that when Abraham Bossé, the mid-seventeenth century artist, painted a series called *The Five Senses*, he used the artichoke to represent the highest refinement of taste.

  Arabian in origin, but cultivated by the Italians, the artichoke is a relatively simple food, but – perhaps because it is difficult to eat – is usually associated with a sophisticated cuisine. It is a delicious food and well worth any effort required in preparation.

  Begin by buying the best and freshest buds you can find. Remember, these are flower buds and you want to buy them when they are tightly closed. Make sure they are fresh. Shrivelled, brown leaves on top should warn you off; look for firm globes with crisp outer leaves that snap when you bend them. Check around the stem; a little black hole probably means a worm has set up house in the part you want to serve for dinner. Rinse well in running water and cut the woody stalk from the base of the bud. With a sharp knife, trim off the prickly top third of the leaves, place upright in a container and cook. A squeeze of lemon juice over each bud adds flavour and helps fix the colour.

You can steam artichokes, simmer them gently or cook them in a microwave. Whichever method you use, cook until a gentle tug pulls a leaf away from the middle part of the bud. Drain and cool. Gently spread, or push the upper petals of the bud apart and carefully remove all of the spiky 'chock', or middle of the flower. At this point you can either remove all of the leaves and search out the succulent bottom section for a special treat or – my favourite – simply serve the cooked bud on a plate with a small side dish of best quality extra virgin olive oil or homemade mayonnaise.

A bit of advice – do not confuse Jerusalem artichokes with globe artichokes; they have nothing in common, except a place on your table.

ASPARAGUS

Good source of:
- Folic acid, beta-carotene, vitamins C and E, potassium

Health benefits:
- Acts as a mild laxative.
- Stimulates the kidneys and works as a mild diuretic.
- Rich in vitamins known to fight cancer; may contain substances that have antiviral properties.

Facts and tips:
- Asparagus was popular with the Egyptians and Romans, but did not find its way onto the fine tables of France – the culinary capital of the Western world – until the reign of Louis XIV.

  Look for asparagus spears with even green upper stalks that end in white woody segments. If kept wrapped in a damp towel in the bottom of the refrigerator, asparagus will last for two or three days. Before use, wash well in running water to remove dirt and sand, then snap off the lower white area. (Some people remove the toughest outer parts of these pieces and use them when making vegetable stock.) Although steaming is the popular way to cook asparagus, they can also be grilled or fried gently in a little extra virgin olive oil. Ideally, asparagus should be eaten as close to the time picked as possible; best on the same day. For that reason, imported bundles of the stuff – which usually cost as much as all the other fruits and vegetables of the meal combined – are often a bit of a disappointment in both flavour and texture. Lucky those who grow their own!

Caution:
- Asparagus is high in substances known as 'purines', which should be avoided by people suffering from gout. Also, asparagus contains substances which are passed from the body in the urine and give it a strong smell.

AUBERGINE (EGGPLANT)

Contains:
- Potassium and carotene

Health benefits:
- This is one of those foods that does not brim over with nutrients on its own, although it contains a good range of vitamins and minerals, but works well with those that do. It is a 'work-horse' vegetable, providing flavour, interesting colour and texture to dishes when teamed up well with nutrient-rich pasta, tomatoes, garlic and soya mince.

  Some experts believe aubergine contains phytochemicals that may help block the formation of tumours.

Facts and tips:
- This is not truly a vegetable, but a fruit which varies in shape and colour across the many varieties of this plant. Eggplant, aubergine's other modern name, was once used in Europe to describe its decorative shape; however, it is used now only in the United States.

  Aubergine is native to Asia and was introduced into Europe by traders during the Middle Ages. Its medical benefits are controversial – in certain parts of Asia it is eaten to treat stomach cancer and viral diseases.

  Aubergine can absorb large quantities of oil when fried. Salting sliced aubergine and allowing it to drain for several hours decreases its capacity to hold fat. Or try brushing slices with a scant quantity of oil and bake in a hot oven until brown.

Caution:
- Aubergines are members of the nightshade family of plants and some people may be sensitive to them. Some evidence suggests this group of plants may increase the discomfort caused by arthritis.

AVOCADO PEARS

Rich source of:
- Monounsaturated fats, vitamin E, potassium, magnesium and phosphate

Also contain:
- Riboflavin, carotene, niacin, vitamin B_6, folate, pantothenate, biotin and vitamin C, iron, iodine, copper, zinc

Health benefits:
- Rich in oleic acid, the monounsaturated fatty acid in olive oil that scientists believe helps maintain a healthy level of cholesterol circulating in the blood.
- Also rich in substances, including antioxidants, that appear to fight cancer.

Facts and tips:
- Avocados originated in South and Central America, and were first recorded by the Spanish explorers during the sixteenth century; however, it took 400 years for them to become popular. Today, they are grown in many temperate areas, including the United States, Mexico and Israel. In Britain during the seventeenth century avocados were known as 'midshipman's butter' because of the oily, spreadable characteristics of their ripe flesh.

  Avocados ripen after being picked. Buy firm, unripe fruit well in advance of when they are to be used and ripen them at home by leaving them in a warm place. Avoid pears with obvious bruising and those which have soft patches.

BANANAS

Rich source of:
- Carbohydrate, potassium, magnesium, iodine, vitamin C and folate

Also contain:
- Carotene, most B vitamins, vitamin E and the amino acid tryptophan

Health benefits:
- Bananas act as a natural antacid and have been shown to help control ulcer pain.
- Many athletes eat bananas during competition as a natural source of quick energy and a means of replacing the potassium lost through exercise and sweat.
- Some diuretics increase the body's loss of potassium; bananas are a rich source of this mineral and a natural way to replace the loss.
- Bananas help reduce the level of harmful bacteria in the gut, especially when eaten with plain yogurt.

Facts and tips:
- Bananas bruise easily and may become unappetizing. Buy firm, slightly green fruit and let them ripen at home. As the banana ripens, the rather dull-tasting complex carbohydrate they contain is transformed into sweet creamy flesh. This shows from the outside, because the skins turn from green to golden yellow; brown flecks on the skin indicate an increasing level of the natural sugars, which are easy to digest.

  Watch out – bananas can cause flatulence, or wind, if eaten before they are fully ripe.

Caution:
- People taking MAOI (monoamine oxidase inhibitor) antidepressants may experience a rapid and dangerous increase in blood pressure, known as a hypertensive crisis, after eating bananas.

BARLEY
See Grain

BASIL
See Herbs

BAY (BAY LAUREL)
See Herbs

BEAN SPROUTS
See Sprouted Seeds

BEANS

An inexpensive, low-fat and versatile food high in fibre and rich in protein, beans are an important part of a balanced diet. They are also rich in complex carbohydrate, minerals (iron, magnesium, manganese, potassium, phosphorous and zinc), and the B vitamins (except B_{12}). This natural combination of nutrients in beans is thought to help control blood cholesterol and sugar levels. Like all high-fibre foods, beans speed foods through the digestive system, thereby eliminating cancer-causing chemicals before they have an opportunity to have an effect.

Facts and tips:
- In general, beans are a good source of protein, but lack lysine and other essential amino acids. For balanced protein, combine beans and grain in the same meal. Beans on toast is a good foundation for a healthy meal.

Warning:
- Kidney beans contain a substance that cannot be digested and can cause severe stomach pain if not removed before eating. Soak kidney beans overnight, boil for 15 minutes, discard and replace water, and simmer for an hour or until soft.

  A small percentage of the population is allergic to a substance found in soya beans and products made from them.

Popular beans are:

aduki beans – red, sweet flavoured beans used in oriental food
black beans – popular in South America and Asia
black-eyed peas – popular in the southern parts of the United States; rich in minerals and most B vitamins
borlotti beans – popular in Italian cooking

broad beans – also known as 'fava beans', these pulses are an important food around the Mediterranean and are a source of folate, vitamin E, manganese, iron and zinc. They can be eaten raw or cooked, or cooked from dried. Low in fat and high in protein, the soluble fibre in these beans helps remove cholesterol from the gut. Broad beans also contain natural antioxidants and phosphorous.

Whereas kidney, scarlet, runner, butter and haricot beans all originated in the New World, broad beans have been feeding Europeans since the Bronze Age. If you plan to enjoy the bean in its pod, the earlier they are picked and prepared the better.

Broad beans are a good source of protein; when combined in food with rice or pasta, they provide all of the amino acids required for normal growth and good health, and also contain iron, phosphorous, niacin, beta-carotene and vitamins C, A and E.

Caution: People taking MAOI (monoamine oxidase inhibitors) antidepressants should avoid broad beans as they contain chemicals which may produce a dangerous rise in blood pressure. Also, some people living in countries bordering the Mediterranean Sea may suffer from favism, a genetic illness causing severe anaemia following ingestion of vicine, a natural substance found in broad beans.

cannelline beans – popular in Italian cooking

chickpeas (garbanzo beans) – the primary ingredient in hummus. Shaped rather like a hazelnut (or filbert nut) with a pointed end, chickpeas are a nutrient-rich legume popular throughout Asia, the Middle East and Mediterranean countries. The food is so ancient it is claimed one of Cicero's family had a wart resembling a chickpea, a facial flaw from which he acquired his name. They are a reasonable source of manganese, vitamin E, folate and iron.

French beans, or flageolet beans – small, elongated green beans popular in France. Eaten fresh or cooked from dried, they add minerals and some B vitamins to the diet.

Although we associate the various forms of this bean – French beans, flageolet and haricots – with French cooking, they are in fact a plant native to the Americas. The vegetable is in fact a pod and its seeds. When the pod, or haulm, is newly formed and its seeds are almost invisible, it is served topped and tailed as a 'French bean'. Later, when the seeds are larger, we serve these long green cylinders as 'snap-bean'; and still later, when the pods are tough and dry, the seeds are shelled from the pods and cooked as flageolet. Each form has its own texture and taste, and each has its own nutritional benefit. Where the young pod contains chlorophyll and vitamins, the fully developed bean is rich in carbohydrates and the vitamins and minerals stored away to support the growth of a future generation of plants.

kidney beans – see above

lima beans – American broad beans containing useful amounts of manganese, folate, zinc and vitamin E

navy beans – popular in the United States, used to make baked beans. Also called 'white kidney beans'. Similar to Italian cannelline beans. High in fibre, medical researchers believe these beans help lower cholesterol levels and fight heart disease and cancer

mung beans – small green beans most often used as sprouts *(see page 32–3)*; contain folate, iron, magnesium and manganese

pinto beans – popular in the United States; an excellent source of dietary fibre

soya (soy) beans – the original home of the protein-rich soya bean is China, where experts believe they were cultivated a millennium before Christ. By weight, soya beans provide more dietary protein than almost any other living thing, vegetable or animal.

There is a significant difference in the rates of certain illnesses, such as breast cancer, among Western and Asian populations: the miracle ingredients in soya beans may be the answer. Eating more soya beans and soya products throughout life may help prevent both breast and prostate cancer. This ancient food has received considerable recent scientific attention because of the combination of medically active phytochemicals it contains. As little as 6 grams (2 oz) of soya protein a day may provide enough of these substances to prevent cancer. They are:

genistein – a substance that appears to stop the spread of cancer in its earliest stages

isoflavones – compounds similar to human steroid hormones that may help control the growth of cancers stimulated by the human forms they mimic. There is some evidence isoflavones also help prevent the negative effects of menopause such as hot flushes (or hot flashes, as they are known in America).

protease inhibitors – naturally occurring enzymes that appear to stimulate the abnormal growth of cells leading to cancer. Substances in soya block the action of these enzymes.

phytic acid – a substance in many plants that block the growth of tumours under experimental conditions. (This is a controversial substance because it is known to bind minerals such as zinc and iron, making them more difficult to absorb from food. Here is a good illustration of the fact that substances in foods may have both a beneficial and negative influence. To be most effective, all nutrients should be consumed in a wide variety of foods to achieve a natural balance.)

Soya is consumed in many forms. Among the most popular are:

beans, excellent in Tex-Mex dishes

tofu, a versatile meat alternative

miso, a fermented paste made from soya beans, grain – barley or wheat – and salt, which is an excellent base for soups, dips and sauces

sauce, brewed by mixing the spores of a mould, *Aspergillius*, with wheat and roasted soya beans, which has been used in Western cooking since the time of Louis XIV of France

protein isolate – about 90 per cent protein, this versatile product has been shown to have a cholesterol-lowering effect when used regularly in the diet

TSP (texturized soya protein), a good meat substitute which is naturally low in fat and calories, but contains significant quantities of protein, calcium, iron, zinc and protective isoflavones

(For more about soya, see also pages 12 and 13.)

BEER
See Miscellaneous Foods

BEETROOT (BEETS)

Good source of:
- Potassium, folic acid, vitamin C

Health benefits:
- Experts believe beetroot is a good detoxifier, helping clean liver and kidneys.
- Food folklore from Eastern Europe claims beetroot fights cancer, but there are no scientific studies to support these claims.

Facts and tips:
- Beetroot may turn urine pink. Some people inherit an inability to metabolize betacyanim, the red pigment in beetroot, and excrete it harmlessly in their urine.

  Many people also enjoy the flavour of green leafy beetroot tops. These are rich in iron, calcium and beta-carotene, which is known to fight the effects of ageing and strengthen the immune system.

BLACK PEPPER
See Spices

BLACKBERRIES

Good source of:
- Vitamin C and folic acid

Health benefits:
- Vitamin C helps strengthen the immune system and fight

infection; it also acts as a powerful antioxidant and blocks the damaging effects of free radicals.

Facts and tips:

- Strewn alongside roads and country lanes, in late summer the thorn-covered briars of blackberries glow garnet with ripe fruit. Not too many years ago, family outings often included picking blackberries for homemade pies and jam. Today, the temptation to take advantage of this bounty is just as great, but should be avoided; heavy metals and other noxious substances in exhaust fumes lie on the fruit and are impossible to remove. Buy cultivated berries from known sources.

  (About jams and jellies – heat-stable nutrients, including minerals, some of the B vitamins and phytochemicals, survive the cooking process. Vitamin C, however, is lost in significant amounts. The pectin used to set some products is a healthy form of soluble fibre. The high sugar content is a good source of quick energy, but may harm the teeth and add unwanted calories to the diet.)

  Like other berries containing red pigment, blackberries contain substances which appear to fight infections, particularly of the urinary tract. In herbal medicine, tea made from dried blackberry leaves is used to treat stomachache, chest congestion and diarrhoea.

Caution:

- Blackberries contain a substance similar to aspirin; people sensitive to aspirin may have a mild reaction to this fruit.

BLACKCURRANTS

Good source of:

- Vitamin C and beta-carotene

Also contain:

- Vitamin E, biotin, thiamin, niacin

Health benefits:

- Exceptionally rich in vitamin C, blackcurrants help strengthen the immune system and provide a good supply of natural antioxidants. By weight, they contain four times more vitamin C than fresh oranges.

Anthocyanins, pigments found in blackcurrant skins, suppress inflammation and inhibit the growth of *E. Coli*, a common type of bacteria responsible for causing a high proportion of diarrhoea and gastroenteritis.

Facts and tips:
- One of the great treasures in Nature's medicine chest, the healing benefits of this plant were well established by the fourteenth century. Blackcurrant is used in syrups and lozenges to treat the inflammation of a sore throat.

BLUEBERRIES

Good source of:
- Energy from natural sugar and modest levels of vitamin C; blueberries are especially useful because of the healing phytochemicals they contain

Health benefits:
- Like blackcurrants, blueberries contain anthocyanins, which fight bacterial infections of the gut and bladder.

  Research suggests blueberries also contain a substance which acts on the walls of the bladder and stops bacteria taking hold on its delicate inner lining. Cranberries contain a similar substance.

Facts and tips:
- Early studies suggest phytochemicals in blueberries actively fight glaucoma, cataracts and other disorders leading to impaired vision and blindness.

BORAGE
See Herbs

BRAN
See Grains

BREAD
See Miscellaneous Foods

BROCCOLI

Good source of:
- Vitamins C and A

Also contains:
- Potassium, iron and folic acid, and is rich in chlorophyll

Health benefits:
- Broccoli is a *Brassica* and therefore a member of the *Cruciferae* family of plants. Like other members of this group, broccoli contains natural chemicals which have been shown to help fight certain cancers (particularly those of the digestive system, lungs and prostate) and block damage to the genetic materials in cells. The dark-coloured florets may be green, blue-green or purple.

Facts and tips:
- Steam, microwave or stir-fry broccoli to preserve its vitamin C; boiling destroys about half of this vital nutrient.

BRUSSELS SPROUTS

Good source of:
- Beta-carotene, folate, antioxidant vitamins C and E, potassium and fibre

Health benefits:
- Another cruciferous vegetable with massive health benefits! Brussels sprouts contain sulphoraphane, which stimulates anti-cancer enzymes, and indoles, natural chemicals that block oestrogen-sensitive cancers.

  The high level of antioxidants and cancer-fighting substances helps ward off cancer, heart disease and damage created by free radicals.

  The high levels of folic acid are beneficial to pregnant women and may help protect people who are susceptible to lung cancer.

Facts and tips:
- Shaded by a top-knot of large leaves, these tiny cabbage-like leaf clusters grow in tight bunches surrounding the thick stalk of the plant. As the name implies, this vegetable appears to have been cultivated around Belgium, where they are mentioned in market records as far back as the early thirteenth century. Over the ensuing centuries they have enjoyed mixed enthusiasm in Europe. However, it was during his period as Ambassador to France that Thomas Jefferson first discovered this fine vegetable and took seeds back to plant in the gardens of his home in Monticello

  Buy firm, bright green Brussels sprouts and avoid those that are yellowed or feel spongy when pressed between the fingers. Cook quickly and without a lid to avoid a build up of unpleasant smelling gases released by the heat. Overcooking breaks down protective indoles.

Caution:
- Brussels sprouts may cause wind.

BUCKWHEAT
See Grains

BULGHAR (BULGUR)
See Grains

CABBAGE

Good source of:
- Potassium, thiamine, folic acid, beta-carotene, fat-soluble vitamins C, K and E, and fibre

Health benefits:
- Helps reduce the risk of constipation.
- Is thought to contain a chemical which aids the healing of stomach ulcers (S-methylmethionine) and others which suppress the formation of tumours in the lower bowel.
- Helps fight infection by stimulating the immune system. Red cabbage is thought to contain substances that are natural antiseptics.

- Research suggests substances in cabbage hasten the breakdown of oestrogen, thus giving some protection against cancers of the womb and breast.
- Has been used to help heal stomach ulcers and in some people relieves the symptoms of colitis.
- Anti-inflammatory.
- Mildly diuretic. Can be used to help detoxify the body.

Facts and tips:
- Cabbage is an ancient vegetable. Although tough to eat and foul smelling, forms of it grew wild and were part of the diet in northern France and England centuries before the Romans brought cultivated varieties with them to propagate and help secure their winter rations. Possibly the Romans also knew of its medical properties. Much later, when Ireland was being settled, the people there took cabbage as their own and grew more there than in any other part of Europe.

  The lowly cabbage has much to offer the health-conscious cook and should appear on the table at least three times a week. Delicious in soup, steamed, stuffed and served as sauerkraut, it is one of our most versatile vegetables.

  Most of the nutrients are contained in the dark outer leaves and much of the valuable vitamin C is lost during cooking, so serve raw, in coleslaw for example, or microwave to cook. Microwaving has been shown to reduce the loss of vitamin C.

Caution:
- Cabbage and cabbage juice can cause wind.

CANTALOUPE

Good source of:
- Water, vitamin C, potassium, beta-carotene

Health benefits:
- Low in calories, but contains substances which may help fight cancer.

Facts and tips:
- The stem end of the cantaloupe should have a smell of the

taste of the fruit; check this and the sensation of weight when you are buying one. When lifted in one hand, a feeling that the fruit is heavy for its size suggests it is fresh and filled with tasty juice.

CARAWAY
See Spices

CARROTS

Good source of:
- Beta-carotene, both soluble and insoluble fibre, vitamin A

Health benefits:
- Populations eating regular, high levels of naturally occurring beta-carotene have lower levels of cancer, especially those of the pancreas and lung, heart disease, cataracts and other conditions related to ageing and damage by free radicals.
- Soluble fibre has been shown to reduce blood cholesterol levels, thus reducing the risk of stroke and atherosclerosis.
- Fibre helps prevent constipation.
- Vitamin A is needed for normal vision; vitamin deficiency first appears as night blindness.

Facts and tips:
- The culinary history of carrots begins well before the birth of Christ, when the edible root, both small and pale at that time, was found by the Romans growing around the villages of Robenhausen, in what is now Switzerland. However, food writer and historian Jane Grigson believes the deep-yellow varieties we enjoy today did not come from northern Europe, but from seeds of purple carrots grown in Afghanistan about seven centuries later. Seeds of this colourful food and a yellow mutant were carried by the Moors to North Africa and Europe, where they spread from Spain to Holland to France and finally to England. The feathery leaves of carrot were much admired and during the Stuart period in England ladies pinned them to their hats and sleeves. Over the centuries carrots have been used to cleanse the digestive tract and – with sugar – prescribed as a mild aphrodisiac.

Enjoy carrots both raw and cooked. Cooking breaks down the tough cells in this root vegetable and releases nutrients from within their indigestible fibre membranes.

Carrots are naturally sweet and make a nutritious food for young children and convalescents because of their pleasant taste. Puréed carrots make good baby food and have been shown to help control diarrhoea; a piece of moist carrot cake (made with ground walnuts) makes a tasty treat for someone who needs something a bit tempting to eat.

Caution:
- Peel carrots. Pesticides and fertilizers containing dangerous organophosphates may remain as residue on the outer surface.

CASHEW NUTS
See Nuts

CAULIFLOWER

Good source of:
- Vitamin C, folate, potassium and fibre; also contains a range of other vitamins and minerals

Health benefits:
- Contains sulphoraphrane, which is thought to help prevent cancer.
- Fibre is good for the bowel and vitamin C is a powerful antioxidant.
- Folate helps prevent spina bifida.

Facts and tips:
- The home of the cauliflower is said to be Turkey, where the plants were cultivated by the Arabs during the Middle Ages. It was introduced into Europe during the sixteenth century by Louis XIV.

  Cauliflower is a member of the same plant family as broccoli and cabbage; it helps reduce the risk of cancers of the colon and stomach. Like other cruciferous vegetables, cauliflower contains sulphur; when cooking, this may collect in the pan and cause an unpleasant taste and smell. Cook rapidly, uncovered. Small florets of cauliflower cook well in the microwave.

Caution:
- May cause flatulence. Flavouring with cumin, ground coriander (cilanho) or tarragon helps digestion.

CELERIAC

Good source of:
- Soluble fibre, vitamin C, potassium

Health benefits:
- Good source of natural antioxidants and fibre, which helps reduce risks from bowel cancer and high blood cholesterol.

Facts and tips:
- A special form of the celery family grown for its root, this delicious, inexpensive, easy to digest, low-calorie vegetable is popular in France and Germany, but ignored by most other cuisines. Try celeriac the French way: cut into very fine strips, boil until just tender, drain and cool, toss with mayonnaise, flavour with mustard or cumin. Celeriac is excellent mashed or cut into thin rounds and deep-fried.

CELERY

Good source of:
- Sodium and potassium

Health benefits:
- Contains 3 n-butylphthalide, a substance known to reduce blood pressure in experimental animals. This fits with the fact that celery was used by the ancient oriental healers to treat high blood pressure! Including celery in the diet once or twice a week may have some possible benefit in controlling high blood pressure and preventing stroke.
- 3 n-butylphthalide may also act as a mild sedative and reduce inflammation.

Caution:

- Celery stores nitrates, which are converted during digestion into nitrates and nitrosamines, which are thought to be carcinogenic in large quantities.

CELERY SEEDS

The seeds have a taste similar to fennel, and are delicious in soups and stews. Celery salt adds flavour to fish and vegetable dishes; make your own by grinding together salt and seeds which have been lightly toasted in the oven.

CHARD (SWISS CHARD)

Good source of:

- Excellent source of carotenes and folate.

Health benefits:

- Adequate folate is important in a mother's diet to promote the normal growth of her infant's brain and spinal cord.

Facts and tips:

- Chard is a form of beet grown for its fine white stalks and green leaves, rather than its root. The stalks can be chopped and used as a crisp, tasty addition to salads or cooked on their own, while the green leaves, which are coarser than spinach and hold their shape better during cooking, are delicious on their own or, cut into fine portions, present an excellent addition to soups, casseroles and quiches.

CHERRIES

Good source of:

- Fructose, a fruit sugar

Also contain:

- Minerals and some vitamin C

Health benefits:

- Contain phytochemicals that may block carcinogens.

Facts and tips:
- This delicious and attractive fruit contains a modest and uninteresting range of nutrients, with one exception: cherries contain ellagic acid, a plant chemical that appears to block the action of natural and synthetic substances that stimulate the development of cancer.

CHERVIL
See Herbs

CHICORY (ENDIVE)

Good source of:
- Folic acid

Health benefits:
- Low-calorie food containing a blend of nutrients.

Facts and tips:
- In France, this plant is called *chicroée frisée*, or 'curly chicory', because of its frizzy leaves ranging from light yellow to dark green in colour. To add to the confusion, many British cooks simply call it *frisée*. Enjoyed by the Egyptians as a winter plant.

CHILLI (PEPPERS)

Good source of:
- Beta-carotene (learn to love chilli for this alone!) and vitamin C

Health benefits:
- Rich in protective antioxidants that help prevent cancer, heart disease and the ageing effects of free radical damage.
- Contains capsaicin, a chemical that has been shown to block pain.
- Some research suggests substances in chillies may help the symptoms of the common cold.

Facts and tips:
- Discovered and named by Christopher Columbus, these peppers were one of the first botanical immigrants from the New World to Europe.
- The hot substances in chilli are thought to have no effect on normal tissue, but they may intensify discomfort from haemorrhoids and ulcers.

CHIVES
See Herbs

CHOCOLATE
See Miscellaneous Foods

CINNAMON
See Spices

CORIANDER (CILANHO)
See Herbs

COURGETTE (ZUCCHINI)

Good source of:
- Pure water, carotene and vitamin C.

Health benefits:
- Contains natural antioxidants that help protect against ageing and degenerative diseases well known in Westernized cultures.

Facts and tips:
- This delicate vegetable was a well-kept Italian secret until the middle part of the twentieth century, when it found its way to the rest of Europe. Seeds taken to America by Italian immigrants made this a popular garden and market food.

CRANBERRIES

Good source of:
- Vitamin C

Health benefits:
- Cranberries contain natural antibiotics which help control bladder infections. The alkaline character of cranberry juice, as opposed to the acidic nature of orange and lemon juice, helps control illnesses involving the digestive and urinary tracts.

Facts and tips:
- Although these tart red berries grow best in the mashy fields of New England, their distinctive flavour and medical benefits are making them a popular fruit in Europe.
- Two glasses of pure cranberry juice a day may help control urinary infections. (Warning: If you think you have an infection and it persists, see your doctor.)

CRESS
See Herbs

CUCUMBER

Good source of:
- A blend of vitamins and minerals, including biotin, iodine and manganese

Health benefits:
- Low in calories.
- Contains sterols, chemicals which have been shown to lower blood cholesterol in experimental animals.

Facts and tips:
- Cucumbers may be the oldest known cultivated vegetable, stretching back before the time of the Romans. One great citizen of Mesopotamia so loved this cooling vegetable-fruit (for it is a fruit) he had a temple built to a god to protect his garden where they grew. It is also claimed that the Roman Emperor Tabors enjoyed this food so much he had

cucumbers raised in earth-filled carts, exposed to the sun during the day and wheeled under frames when the weather was too cold or threatening.

Most of the goodness in cucumbers is in the skin. To get maximum benefit, buy unwaxed varieties.

CUMIN SEEDS
See Spices

CURRY
See Spices

DANDELIONS

Good source of:
- The leaves contain more iron than an equal weight of spinach, plus vitamins A and C, and potassium

Also contain:
- Beta-carotene, most B vitamins and very small amounts of essential fatty acids

Health benefits:
- Dandelions are rich in phytochemicals with medicinal value. Substances in the leaves are thought to help purify the liver.
- If the roots are properly prepared, they are said to help clear debris from the spleen, gallbladder, pancreas and kidneys. Dandelion root coffee is sometimes prescribed as a natural means of treating hypoglycaemia because it contains an insulin-like substance. Tea made from the root is also used to benefit adult-onset diabetes and help lower blood pressure.
- Fresh juice or a powered form of dandelion is used in China to treat infections of the upper respiratory tract and digestive system.

Facts and tips:
- The Russians so love dandelions they affectionately refer to their children with the same word. Called *pissenlit* (wet the bed) in French as a reminder of its strong diuretic action,

dandelion is a good winter salad plant, particularly when mixed with another wild plant, sorrel.

DATES

Good source of:
- Calcium, phosphorous, magnesium, vitamins B_1 and B_2; packed with natural sugar for quick energy

Health benefits:
- Dates provide a healthy balance of most necessary micro-nutrients.

Facts and tips:
- Dried dates and foods containing dates are excellent fare for munching on long trips.

EXOTIC FRUITS

Good source of:
- Vitamin C, carotenes, natural fruit sugars

Health benefits:
- Contain natural substances that help strengthen the immune system and fight degenerative illnesses.

Facts and tips:
- The modern world trade in food brings the tastes and textures of fruits from all over the world to our local supermarket. Although the price is dear and no fruit that has been shipped thousands of miles ever has the full taste of that enjoyed in its native environment, a dessert or special breakfast treat made of one or more exotic fruits reminds us just how exciting foods can be. Fresh dessert ices made with the flesh of exotic fruits are particularly memorable. Or try adding some of the more highly scented exotic fruits to freshly squeezed orange juice. Some of the most popular exotic fruits are:

Cape gooseberries (also known as Chinese lanterns) – Nature encloses these little orange beauties in a paper-like wrapping which splits open when they are ripe. Packed with beta-carotene and vitamin C, they make a beautiful garnish with fish or lemon cake, or as topping on a mixed fruit salad.

Kumquat – Tiny citrus fruits from China packed with vitamin C and encased in a skin so tender it can be eaten along with the fleshy interior. Lovely whole or sliced.

Logans – Another Asian fruit, the logan has a hard shell which needs to be removed. Served fresh, it is rich in vitamin C.

Loquats – Sweet and tender, the flesh of this oriental fruit resembles a blend of cherries and plum. High in beta-carotene, loquats present a mystery because they contain no vitamin C.

Lychees – Most people recognize these from the canned varities served as dessert in many Chinese restaurants. However, try the fresh fruit; it is packed with vitamin C and has a wonderful light aroma of flowers.

Passionfruit – From South America, this somewhat difficult to handle fruit is loaded with seeds and has earned the name 'little pomegranate'. The taste is exquisite.

Persimmon – Originally from Japan but now grown in many temperate parts of the world, the persimmon, or Sharon fruit, needs to be eaten very ripe or the effect on the mouth is very unpleasant. However, resist cutting the fruit until the colour is deep orange and the tough outer skin gives slightly to gentle pressure, and you have a real treat in store. Cut open the fruit, spoon out the soft orange flesh into a fruit salad or onto a dollop of fromage frais, and, as you increase your daily intake of vitamin C, you will enjoy a most wonderful treat.

FENNEL
See Herbs

FENUGREEK
See Spices

Good source of:
- Fibre, fruit sugars, folate, potassium, calcium, phosphorous, iron, copper, manganese and zinc

Health benefits:
- The minerals help maintain the normal chemical balance in the body and are essential for strong bones.

Facts and tips:
- Excellent high-energy snack food. A good food choice for growing children and women concerned about osteoporosis.

GARLIC

Sometime listed as a herb, garlic is one of the most valuable plants in the garden and deserves a listing on its own. While its distinctive flavour and aroma have been prized by cooks in almost every culinary culture since the time of the Pharaohs, garlic has also gained respect for its remarkable health benefits.

Good source of:
- Selenium and other minerals, powerful associate nutrients, including diallyl sulphide (DAS), which has been shown in laboratory studies to slow tumour growth and deactivate cancer-causing substances. In particular, it may be effective against the carcinogenic action of nitrosamine.
  Garlic powder contains some tryptophan.

Health benefits:
- As a member of the *Allium* family, garlic contains allyl, a natural antiseptic believed to protect against common infections and other illnesses as dangerous as dysentery and typhoid.
- Research results suggest chemicals in garlic help lower blood cholesterol levels, thus reducing the risk of heart disease. It also is said to help control hypertension, or high blood pressure.

Facts and tips:
- If garlic makes you burp, or gives you gas, the problem may be found in the growing green germ found in older cloves. Try using some garlic with these removed to see if the problem disappears. Simply cut the clove in half, locate the green embryonic plant at its base and remove it with the point of a knife. Many people find this also removes much of the overly pungent smell and flavour found in mature garlic.

  Gardeners may know garlic for its insecticidal properties; grown in vegetable and fruit gardens, it discourages aphids.

GINGER

Good source of:
- Magnesium, zinc and the B vitamins

Health benefits:
- Contains phytochemicals that calm nausea and prevent travel sickness.

Facts and tips:
- Gingersnaps are a good snack for children during a long trip.
- A nibble of candied ginger after a rich meal can help calm the stomach and aid digestion.

GRAPEFRUIT

Good source of:
- Pectin (an insoluble fibre) and vitamin C

Health benefits:
- Insoluble fibre helps reduce blood levels of LDL-cholesterol, which is thought to be the culprit in atherosclerosis.
- All citrus fruits contain the phytochemical limonene, which may have cancer-fighting properties.

Facts and tips:
- Recent research suggests there is a negative interaction between grapefruit and certain medications, particularly calcium-blockers used to treat high blood pressure. If you have any questions, ask your doctor.

GRAPES

Good source of:
- Potassium, boron, magnesium, copper and some iodine

Health benefits:
- Boron (a trace mineral) may help post-menopausal women avoid osteoporosis by maintaining a higher level of blood oestrogens. .
- All grapes help clean the liver, intestines and kidneys. They also contain ellagic acid, which is thought to have anti-cancer benefits.
- Black grapes contain resveratrol, a phytochemical believed to prevent atherosclerosis in laboratory animals.

Facts and tips:
- The only way to judge the sweetness of grapes is by trying one. (*See also* Wine.)

GREEN TEA
See Miscellaneous Foods

HAZELNUTS
See Nuts and Seeds

KALE

Good source of:
- Folate, vitamin C, beta-carotene, iron and calcium

Health benefits:
- Kale is a rich source of a form of calcium easily absorbed by the body. The calcium in many other foods is not readily available to the body during digestion.
- Kale and other members of the cabbage family contain a powerful phytochemical, sulphoraphane, that is thought to block the action of some cancer-causing substances. The same group of vegetables also contain substances called indoles which stimulate the liver to break down excessive

amounts of circulating sex hormone needed for the growth of certain forms of cancer; breast cancer is an example.

Facts and tips:
- A member of the cabbage family, kale has a distinctive aroma while cooking which some may find rather strong. However, its high vitamin, mineral and natural antioxidant content make it a worthy addition to your menu.

  Medical experts at the Harvard University School of Public Health have specifically cited kale as a food that fights cancer.

KIWI FRUIT

An exotic fruit that deserves its own listing!

Excellent source of:
- Vitamin C and potassium

Health benefits:
- Kiwi fruit is thought to help lower high blood cholesterol because of its soluble fibre content.
- Vitamin C is a powerful antioxidant thought to help reduce risk from cancer and heart disease.

Facts and tips:
- First grown in China, where it was known as 'the Chinese gooseberry', kiwi fruit is now most often associated with New Zealand, where it received its name because of its resemblance to the body of the flightless bird the kiwi.

  Why not substitute a kiwi fruit for orange juice at breakfast? Simply place one in an egg cup, slice off the top and scoop out the delicate flesh with a small spoon. Kiwi fruit is sweetest when the fruit is ripe and soft.

KOHLRABI

Good source of:
- Vitamins C and E, potassium.

Health benefits:
- Rich source of indoles, the plant chemicals which help reduce the risk of breast cancer. Also contains isothiocyanates, thought to block cancer of the lower bowel.

Facts and tips:
- Shredded kohlrabi is a good addition to slaw and salads. Buy small, young bulbs which are still tender.

KOMBUCHA

A possible food for the future.

Good source of:
- B vitamins, amino acids and phytochemicals

Health benefits:
- Drinking tea fermented by kombucha is thought to strengthen the immune system and provide the vitamins and proteins needed for rebuilding body tissues.

    Health claims are extensive, and range from asthma and liver aliments to cancer, but systematic documentation is needed.

Facts and tips:
- Kombucha is a harmonious community of bacteria and yeast rather than a true fungus, which it resembles. It grows rapidly on a mixture of tea and sugar, producing essential amino acids and vitamins as part of its life processes.

    Kombucha's history can be traced back to Manchuria, from where it was taken to Japan about 2,000 years ago. It was then carried to what is now Russia, where it became well known as a home remedy. Solzhenitsyn, the dissident Soviet poet, claims it saved his life during imprisonment.

LEEKS

Good source of:
- Potassium and folate

Health benefits:

- Potassium plays an important role in maintaining normal blood pressure levels. Leeks help balance the high levels of sodium found in modern Western diets. Potassium also assists normal kidney function.

Facts and tips:

- Nero is said to have eaten leeks to improve his voice. More recently, claimed as a symbol of Wales and worn pinned to the hats of the Welsh Regiment on St David's Day, leeks have been credited with healing properties by healers there and in the other parts of the island nations off the coast of France. For example, in the medieval gardens in the grounds of the Sherburn hospital, in County Durham, the growing of leeks was entrusted to lepers. It was later said that this terrible disease declined more quickly in Ireland than other countries of Europe because of the vegetables they ate – and also the increased awareness of the benefits of good hygiene!

Warning:

- Leeks can cause flatulence.

LEMONS

Good source of:

- Vitamin C, pectin

Health benefits:

- Useful as a natural antiseptic, astringent and diuretic.
- Contain limonene, a phytochemical thought to block the action of cancer-causing chemicals, and other plant substances with health benefits.

Facts and tips:

- Citrus fruits may be coated with wax or sprayed to retard spoilage. Like all other fruits, wash lemons before using and if you plan to use the rind in your food, scrub well with a soft brush. If possible, in food only use peel from untreated lemons, if you can find them in the shops.

LENTILS

Good source of:
- Protein, complex carbohydrates, both soluble and insoluble fibre, folate, copper, iron

Health benefits:
- Low in fats and high in the other macro-nutrients needed for good health, lentils are a good choice for a modern healthy diet.
- The fibre content helps lower cholesterol and maintain a healthy bowel.
- Phytates in lentils (and other pulses) are believed to help fight cancer and tumour growth.
- The folate and iron content in lentils make them a good choice during pregnancy.

Facts and tips:
- An ancient food mentioned in the Bible.

LETTUCE

Source of:
- Beta-carotene, vitamin C, folate, potassium, iron, some fibre

Health benefits:
- Natural antioxidants help fight degenerative diseases, including cataracts.

Facts and tips:
- 'Lettuce' is a general term covering a wide range of low-growing, green leafy plants. They have been prized for their crispness and the cooling effect of their high-water content for millennia, as attested to by the wall-painting of the Egyptian pyramids. Lettuce's white juices were once said to have a soporific effect, which was recalled when it was claimed to have a disastrous effect on Peter Rabbit's relatives when they dined in Mr McGregor's garden!
  See also Chicory.

LIMES

Good source of:
- Vitamin C and other natural antioxidants and bioflavins

Health benefits:
- *See* Lemons

Facts and tips:
- As explained in Chapter 2, during the mid-eighteenth century sailors and officers aboard ship for long periods contracted a devastating illnesses called scurvy. The British naval surgeon, James Lind, thought their restricted diets during long voyages was the cause of these outbreaks and insisted ample stores of limes be stored aboard the ships.

MANGO

An exotic fruit that deserves mention on its own. Think of the soft and deliciously flavoured flesh of a ripe mango as a packet of natural healing.

Good source of:
- Vitamin C and beta-carotene, fibre, potassium, carbohydrate energy

Health benefits:
- Mangoes contain vitamins and phytochemicals that protect against ageing and the degenerative illnesses of our Western lifestyle.
- Dietary fibre helps fight bowel cancer.

Facts and tips:
- Buy mangoes by touch; the best are firm with a slight 'give' when pressed. Avoid fruit with soft spots and bruises.

MARJORAM
See Herbs

MELONS

Good source of:
- Potassium, vitamin C and carotenoids

Health benefits:
- Don't overlook the importance of water supplied by fruit in a healthy diet. It is free of contamination from chemicals and as fresh as Nature can provide. Melons have a high water content and were prized by ancient travellers as a safe way to quench thirst during long journeys.

  It is thought that melons originated in Asia and the melon seeds discarded by traders on their way west spread the fruit throughout much of the Middle East.

Facts and tips:
- Choose by smell; the stem end should have a sweet, almost floral smell.

MILLET
See Grains

MINT
See Herbs

MOLASSES (BLACKSTRAP MOLASSES)
See Miscellaneous Foods

MUSHROOMS

Good source of:
- Protein, carbohydrate and minerals: potassium, phosphate, iron and sulphur

Also contain:
- Tryptophan

Health benefits:
- Good news! Two delicious types of oriental mushrooms, the Shiitake and Reishi, are packed with cancer-fighting chemicals. Tests show that a compound in Shiitake,

lentinan, strengthens the immune system and fights tumour cells. In Japan, this substance is used to help treat cancer. Both Shiitake and Reishi mushrooms are good sources of B vitamins riboflavin and niacin.

According to Kenneth Jones, author of *Shiitake: The Healing Mushroom*, other mushrooms with anti-tumour properties are oyster, enokitake, nameko and matsutake.

Facts and tips:

- White mushrooms, *Agaricus bisporus*, are sold according to their size and stage of maturity, as button, closed cup and flat. Mushrooms do not contain chlorophyll and are not a true 'plant', therefore they are very dependent on their environment for nutrients. Almost all mushrooms we buy are 'farmed' – even the exotic types, like Shiitake and oyster mushrooms, are produced under artificial conditions. Some are grown on a compost based on dried poultry manure, DMP. Although this is an approved farming method, there is some concern it may introduce some residues from the poultry farming. Organic mushrooms are best. Ask your store manager to get some in if necessary.

  These delicious foods require some common sense. First, *do not* pick your own mushrooms unless you are absolutely sure of what you are doing. Ceps, morel and girolle add deep forest flavours to sauces, egg dishes and risotto, but other varieties are quite deadly. Second, mushrooms can concentrate dangerous heavy metals and other toxic substances from the soil in which they are grown; for example, nitrosamines may be concentrated in the flesh of farmed mushrooms when raised on manure and straw from farms using large amounts of nitrate fertilizers.

  To clean mushrooms, just brush off any woody bits or flecks of compost with a mushroom brush or piece of soft paper kitchen towel. The water mushrooms absorb when washed can alter the way they behave in the pan.

MUSTARD (DRY)
See Spices

MUSTARD (GREENS)

Good source of:
- Beta-carotene, vitamin C, iron, calcium

Health benefits:
- A member of the *Cruciferae* family, mustard is packed with cancer-fighting phytochemicals.

Facts and tips:
- Use young leaves; older foliage becomes tough and has a stronger smell during cooking. *See also* Sprouted Seeds.

NECTARINES

Good source of:
- Vitamin C, carotene, potassium, phosphate, magnesium, fruits sugars and pectin

Health benefits:
- Contain antioxidants that help prevent the breakdown of essential fats.

Facts and tips:
- Do not buy green fruit, it rarely ripens well at home and usually lacks flavour. The best fruit smells best; let your nose lead you to the sweetest fruit.

NETTLES

Good source of:
- Beta-carotene, vitamins A and C, chlorophyll, magnesium and more iron by weight than spinach!

Health benefits:
- Shoots of young nettle leaves, gathered first thing in the spring, were once said 'to purify the blood'. We now know these prickly leaves are packed with vitamins, minerals and associate nutrients that strengthen the immune system and support the production of blood cells.

Facts and tips:
- Nettles are delicious when cooked with spinach or sorrel.
- Pick only the youngest shoots and beware of the sting!

NUTMEG
See Spices

OATS
See Grains

OLIVES

Good source of:
- Vitamin E and other natural antioxidants, and oleic acid, a monounsaturated fat

Health benefits:
- Olive oil, used in all forms of cooking, is high in mono-unsaturated fats which are thought to be healthier for the heart than the saturated varieties found in red meat and butter.
- Olive oil has been used for centuries to treat gallstones, constipation, burns and earache.

Facts and tips:
- Olives are grown for both their distinctive taste and fine oil, which is a distinctive flavour in Mediterranean cooking.

Warning:
- When obtained from the tree, olives are very hard and need to be treated with salty brine before eating. As a result, the edible results are delicious, but have a high sodium content. Because excessive levels of sodium in the diet contribute to high blood pressure, and thereby to heart disease, olives should be enjoyed in moderation.
  See also Oils.

ONIONS (INCLUDING SHALLOTS AND GREEN ONIONS)

Good source of:
- Healthy blend of vitamins and minerals; packed with miracle phytochemicals

Health benefits:
- Research has shown members of the *Allium* family contain substances that act as natural antibiotics, block tumour formation, reduce risk from dangerous blood clot formation, lower blood pressure and raise the level of protective HDL cholesterol.
- Onions may also help control digestive and respiratory problems, and treat the symptoms of colds and 'flu.

Facts and tips:
- Since ancient times onions and other members of the *Allium* family, which includes garlic and leeks, were thought to contain healing and protective substances. Little is known of the true origins of onions as food, but they are mentioned in ancient and modern cuisines throughout the world.

Warning:
- A small percentage of the population experiences migraines after eating members of the *Allium* family.

ORANGES

Good source of:
- Vitamin C, thiamin, potassium and pectin

Health benefits:
- Pectin may help lower blood cholesterol levels.
- As one of the most powerful, natural antioxidants, vitamin C helps fight cancer, heart disease and other degenerative conditions linked with tissue damage caused by free radicals.

Facts and tips:
- If you use orange slices, buy fruit that is not coated with protective wax.

OREGANO
See Marjoram

PAPAYA

A tropical fruit that deserves its own listing!

Good source of:
Beta-carotene, vitamin C, potassium, fibre

Health benefits:
- Papaya contains papin, an enzyme known to break down protein. In fact, it is so powerful, farm workers handling papayas need to protect their hands with gloves. Papin is used by the food industry to tenderize meat and has found medical usage as a means of reducing pain from swollen or ruptured spinal discs.

Facts and tips:
- No papaya tastes or smells like a papaya picked fresh from the field. But if you long for fresh papaya and, like most of us, must settle for one from the local supermarket, be picky before you spend your money. Look for fruit free of all bruising and soft spots. If they are slightly green, all the better. Take one home and shelter it in a warm place, turning daily, until the skin has a pink tint and the fruit gives slightly when gently squeezed. Cut in half, remove the seeds and scoop out the flesh to serve as it is or add to a fruit salad.

PAPRIKA
See Spices

PARSLEY
See Herbs

PARSNIPS

Good source of:
- Complex carbohydrate and fibre. Rich in vitamins E and C, and contains folate

Health benefits:
- High-fibre foods help cleanse the digestive system.
- Members of the *Umbelliferous* vegetables, which also includes carrots, parsley, dill and celery, parsnips contain terpenes, which may reduce the spread of cancer cells and deactivate substances that encourage tumour growth.

Facts and tips:
- Enjoy this sweet and easy to prepare root vegetable as an alternative to potatoes. They can be mashed, baked and fried. Young roots are the sweetest and have the advantage of containing a small central core; as the vegetable matures, this core expands and toughens. If you are using large parsnips, removing the core will improve the texture of the food you prepare.

PEACHES

Good source of:
Vitamin C, selenium, beta-carotene, energy

Health benefits:
- Antioxidants help fight damage created by excess free radicals.

Facts and tips:
- Select by sweet smell and depth of colour; peaches bought when green do not ripen well.

PEARS

Good source of:
- Potassium, folic acid, carbohydrate, natural fibre and vitamin C. Fresh, ripe fruit is a good source of fluid.

Health benefits:
- Pears are low in substances that cause allergies and make a good food for young children.
- They are also a good food choice for those who are ill or convalescing.

Facts and tips:
- Dried pears are a good addition to a back-pack or as a emergency food.

PEAS

Good source of:
- Vitamin C, folate, thiamin (a B vitamin) and phosphorous. Peas contain substantial amounts of protein, carbohydrate and fibre.

Health benefits:
- Carbohydrates are our best energy source.

Facts and tips:
- An ancient vegetable, the pea was most often dried and used as a stored food until the seventeenth century, when the delicate flavour of fresh, young peas was enjoyed in Europe, and later in the United States, where the great gardener and two-term President Thomas Jefferson enjoyed both the appearance of this delicate plant in his garden and the sweetness of its unripened seeds on his table.

  Fresh peas are a food that profits from being frozen. Natural processes in fresh peas begin converting sugars into starch as soon as peas are picked from the vine. Freezing is done very soon after the peas are picked and stops this process. But the long trip from the field to the greengrocer's permits a considerable time for the taste-limiting conversion process to occur.

  Chinese herbalists use peas to aid indigestion, to strengthen the digestive system and to aid normal intestinal flow.

PEPPERMINT
See Mint

PEPPERS AND CHILLI *(CAPSICUM)*

Good source of:
- Beta-carotene and vitamin C. Peppers are packed with natural bioflavonoids and antioxidants. Weight for weight, both red and green peppers contain more vitamin C than oranges.

Health benefits:
- Packed with natural antioxidants that help defend the body against cancer.

Facts and tips:
- Green, yellow and red peppers are all the same vegetable at different stages of ripeness. In some cases, more than one colour will be present on the same fruit. (Peppers are one of those 'vegetables' – like tomatoes and cucumbers – which are really the fruit of a plant.) You may have noticed that red peppers taste sweeter than green peppers; this is because the sugar content increases in the fruit as it matures.

  What we now think of as the native foods of countries on either side of the Mediterranean Sea, India, Pakistan and Eastern Europe, would lose their character if *Capsicum*, or 'box' peppers suddenly disappeared. It is hard to believe that a fruit that now can be found in a spectacular range of shapes, flavours and peppery hotness was unknown outside the Americas until after Columbus's voyage of discovery.

  In the United States, canned red peppers are often called 'pimentos', from their French name, *piment*, which is sometimes reserved for chilli peppers.

PINE NUTS
See Nuts and Seeds

PINEAPPLE

Good source of:
- Potassium and vitamin C

Health benefits:
- Pineapples contain bromelain, an enzyme which breaks down proteins and has been used to help people with problems digesting proteins.
- Other medical applications for bromelain include treatment of heart disease, breaking up blood clots and increasing the effectiveness of certain antibiotics.

Facts and tips:
- When buying, trying pulling a leaf from the centre of the tuft of leaves at the top of the fruit; if a modest tug removes the leaf, the pineapple is ripe. Avoid any fruit with bruising or soft areas.

   A small percentage of the population will develop an allergic reaction to pineapple.

POTATOES

Good source of:
- Complex carbohydrate (starch) and fibre

Also contain:
- Useful amounts of vitamin C, potassium and phosphate

Health benefits:
- The starchy (complex) carbohydrate in potatoes are a good form of slow-release energy that keeps blood sugar levels steady and hunger pangs at bay.

   Potatoes are thought to contain substances which are antiviral and anti-inflammatory. They also contain natural antioxidants and are known to aid conditions of the digestive system, including ulcers and colic.

Facts and tips:
- Potatoes are fattening and contain little nutritional value: true? No, absolutely the opposite. A report in the *British Food Journal* in 1992 set the record straight.[1] Potatoes are rich in fibre, carbohydrate, vitamin C and the minerals potassium and phosphorus. A medium-sized potato, steamed, baked or microwaved, contains about 60 calories; according to the report, you would need to eat 95 130 gram (4.5 oz) potatoes to gain 1 kg (2 lb) of body fat.

   For the occasional treat, if you love what the British call 'chips' and the Americans 'French fries', potatoes deep fried in fresh olive oil are at least neutral in their effect on blood cholesterol levels.

   As consumer tastes shift towards rice and pasta as primary sources of carbohydrate, fresh potatoes have become less popular.

Herbal practitioners in China believe potatoes help heal eczema and inflammation. Their method was to lace raw grated potato over an affected area of the skin and cover it with clean gauze. After two hours, the gauze and potato were removed, the affected area gently cleaned and the process repeated until the redness disappeared. Needless to say, frequent changes were advised.

PRUNES

Good source of:
- Beta-carotene, vitamin B_6, potassium, copper, iron and soluble fibre

Health benefits:
- Soluble fibre helps reduce the blood level of damaging LDL cholesterol.
- The combination of nutrients and associate nutrients in prunes help fight cancer and protect against coronary heart disease.

Facts and tips:
- These dried plums with a high natural sugar content are used in Asia to treat stomach and digestive problems, as well as constipation.

PUMPKIN SEEDS
See Nuts and Seeds

PUMPKINS

Good source of:
- The natural antioxidants carotene, folate and vitamins C and E

Also contain:
- Potassium, calcium, magnesium, phosphate and some iron, copper and zinc

Facts and tips:
- Loved in North America as a traditional ingredient for pies during the winter season and enjoyed in the Caribbean, the Middle East, Australia and most of Europe, pumpkins are grossly underrated in Great Britain, where they are mainly thought of as material for jack o' lanterns and ingredients in ethnic cooking.

QUORN

Good source of:
- Protein and some B vitamins

Health benefits:
- Low-fat protein source.

Facts and tips:
- Do not confuse Quorn with tofu and other soya protein products; Quorn is made from mycelium (part of a fungus) and lacks many of the phytochemicals found in soya.

RADISHES

Good source of:
- Vitamin C; also calcium, copper, iron, phosphorous, potassium, sulphur and phytochemicals

Health benefits:
- Radishes are a mild stimulant to the kidneys and act as a diuretic, helping lower the blood pressure. They are thought to help relieve catarrh and improve the teeth, gums and nails.
- Radish is a member of the *Cruciferae* family and therefore blessed with phytochemicals which research indicates help fight certain forms of cancer.

Facts and tips:
- Radish has been enjoyed as a food since the time of the Pharaohs. Dozens of varieties are raised throughout the world and enjoyed for the crisp, peppery root. Raw radishes should be chewed well before swallowing. Some people find they cause mild indigestion when not well masticated.

As they stimulate salivation, radishes are frequently included in salads or 'nibbles' served during the early part of a meal. The sprouted seeds and young leaves are also excellent additions to salads.

RAISINS

Good source of:
- Carbohydrate, potassium, calcium, magnesium, iron, copper, selenium

Health benefits:
- Iron is needed for haemoglobin, the substance in red blood cells that transports oxygen around the body.
- A good energy source that is low in fat and contains no cholesterol.

Facts and tips:
- Raisins are a 'dried vine fruit'; others are currants and sultanas.

RASPBERRIES

Good source of:
- Vitamin C, potassium, calcium and fibre

Health benefits:
- Red-skinned fruits contain natural antibiotics that help fight infection.
- High in flavour and low in calories, raspberries are a good choice when watching your weight.

Facts and tips:
- The only way to tell if raspberries are sweet is by tasting them.

RICE
See Grains

ROSEMARY
See Herbs

RYE
See Grains

SAFFRON
See Spices

SAGE
See Herbs

SAVORY
See Herbs

SEA KALE

Good source of:
- Beta-carotene, vitamin C, iron, potassium, some B vitamins, fibre

Health benefits:
- Contains phytochemicals that fight cancer, particularly of the lung and colon.

Facts and tips:
- Sea kale is a crucifer, but unrelated to the kale, broccoli and other rather smelly members of the *Brassica* family descending from wild cabbage originating on the sea cliffs. Sea kale was cultivated from plants growing on beaches of the northern coasts of Europe, usually along the edges of the furthest reaches of the tide. It is an English vegetable, which was admired and enjoyed by Thomas Jefferson, who tried hard to popularize it in the United States; unlike many other treats from the gardens of Montechello, however, it never became popular there.

SEA VEGETABLES

Good source of:
- Protein, B vitamins, vitamin K, iron, iodine, magnesium, potassium, copper, zinc

Health benefits:
- A balanced supply of minerals and B vitamins helps stabilize the body's chemistry and strengthens the immune system *(see also Algae)*
- A rich source of iodine, a trace mineral needed for normal health and missing in foods grown in certain regions of the world.

Facts and tips:
- Edible forms of sea vegetables include:

Agag-agag (kanten), a vegetarian alternative to gelatine

Arame (sea oak), rich in calcium and used in miso soup

Kelp (kombu, konbu, laminaria), found around the world; helps to soften pulses when added to the cooking water; popular in Japan, where it is used to wrap sushi

Laver (noir), widely enjoyed in Wales, Ireland and Scotland, laver is also popular in Japan

Samphire (sea asparagus, glasswort), not a true sea vegetable, but similar, this salty treat has become a favourite of many modern chefs.

Wakame, full of chlorophyll, B vitamins and trace minerals, when added to miso or added to other soups or grain dishes makes a powerful contribution to the daily nutrient intake

SHALLOTS
See Onions

SORREL

Good source of:
- Iron, carotenoids, vitamin C, potassium, magnesium, chlorophyll

Health benefits:
- Potassium is needed to maintain the acid-base balance in body fluids.

Facts and tips:
- A tender leaf-plant with a distinctive flavour that makes a delicious quick soup or sauce for fish. Adding a few chopped young leaves gives a good contrast to the taste of other salad greens.

SOYA (SOY) BEANS
See Beans

SPINACH

Good source of:
- Beta-carotene, vitamin C, iron, calcium, folate, potassium and magnesium; also provides a rich supply of chlorophyll.

Health benefits:
- May help lower blood cholesterol; contains substances that block damage from excessive free radicals.

Facts and tips:
- Spinach was cultivated first in Persia, around the sixth century AD, and was so highly prized it was soon considered a dish of great distinction. The food writer Jane Grigson recounts that Emperor T'ai Tsung, of the T'ang period, asked his tribal rulers to send him seeds of the best plants growing in their country; spinach seeds were sent by the King of Nepal. However, it took many centuries for spinach to find its way to European tables, and it was unknown during the times of the great early physicians. As a result, no medical benefits were assigned to this plant until modern times, when it was found to be a rich source of vitamins, calcium and other minerals.

    Where we once insisted children ate great quantities of this vegetable, many now worry that the oxalic acid levels in the leaves may counter some of the beneficial effects of the other nutrients it contains. Spinach – like everything else – should be enjoyed in moderation.

SPROUTED SEEDS

Good source of:
- Vitamin C, B vitamins, protein (amino acids) low in calories

Health benefits:
- Vitamin C is a powerful natural antioxidant.

Facts and tips:
- Life begins in germinating seeds when they are exposed to moisture and warmth. As the carbohydrate, oil and protein nutrients in the seed are utilized to build the first stage, or sprout, of a growing plant, the vitamin level and the amount of certain amino acids present increase over those found in the seed.

  Most people associate sprouts with small green mung beans, but excellent tasting sprouts are also obtained from aduki beans *(see p.92)*, alfalfa seeds *(see p.83)*, chickpeas *(see p.93)* and lentils *(see p.118)*. Chickpea sprouts are tough and need cooking before they are used.

  Seed-leaf or sprouting plants for salads include:

  • *Mustard (and cress)* – Often grown and eaten together, these minerals are vitamin-rich salad foods have been enjoyed since the time of the ancient Greeks.

  • *Mung beans* (most often thought of as 'bean sprouts') – High in energy and 'crunch appeal', bean sprouts are an excellent source of B-complex vitamins and vitamin C. Full of water and bulky, they add few calories but considerable bulk to a salad or stir-fried dish.

STRAWBERRIES

Good source of:
- Vitamin C and fibre

Health benefits:
- In addition to the powerful antioxidant effects of vitamins C and the protective effects of fibre, strawberries also supply ellagic acid, a phytochemical that blocks the action of carcinogens.

Facts and tips:

- The health benefits of strawberries were recognized as far back as the seventeenth century. Castelvetro told his readers to make a 'decoction' of strawberry leaves and roots to cure inflammation of the liver and to regulate the kidneys.

  Let your nose tell you which strawberries to buy. No matter how beautiful the berries look, if they have no smell they will have no taste.

  You may want to think twice about buying irradiated fruit; by extending the shelf-life of the food with irradiation, bacteria and other micro-organisms living on the skin of the fruit are given longer to build up their populations before you eat them.

SUNFLOWER SEEDS
See Nuts and Seeds

SWEET POTATOES

Good source of:

- Carotenoid and phytochemicals

Health benefits:

- *See* Yams

Facts and tips:

- Native to Central America, these delicious root vegetables were taken to Spain by Christopher Columbus on his return from discovering the New World. Here they were propagated, found favour and promptly called 'Spanish potatoes'.

  Sweet and having good texture, sweet potatoes have enjoyed several periods of popularity. In Tudor times they were made into crystallized candy and enjoyed as a supposed aphrodisiac. Josephine set the ladies of Paris talking during Napoleon's reign by serving them at banquets, thus making them a fashionable dinner treat.

  Sweet potatoes are not to be confused with yams, which have a similar shape but lack the delicate colour, delicious sweetness and fine texture of the real thing.

TARRAGON
See Herbs

THYME
See Herbs

TOMATOES

Good source of:
- Natural antioxidants; especially beta-carotene and other carotenoids and vitamins E and C; also potassium

Health benefits:
- Tomatoes and foods rich in tomato concentrate have been associated with a reduced risk of cancer and heart disease in men. (This may be due to the presence of lycopene, a carotenoid.)
- In folk medicine, tomatoes were used to treat constipation, kidney disease and liver trouble.

Facts and tips:
- When we think of tomatoes as being part of a healthy Mediterranean diet we may be misled into thinking they are native to the region. In fact, they were imported from the Americas to Europe, where they were first considered to be a curiosity and then a potential health hazard, possibly causing excessive lust!

  Buy tomatoes that have good full colour; they taste best. When possible, stake out some plants in your backyard and enjoy a feast of sweet-smelling, sun-drenched fruit in the late summer.

  Members of the nightshade family of plants, tomatoes may cause sensitivity in a very small percentage of the population.

TURNIPS

Good source of:
- Fibre, starchy carbohydrate, folate, vitamin C, some B vitamins

Health benefits:
- A cruciferous vegetable, turnips contain phytochemicals that fight cancer. One group of these, the glucosinolates, are destroyed during cooking, so eat your turnips young and raw for best effect. Rutabagas, a member of the turnip family, is particularly high in anti-cancer substances.

Facts and tips:
- Turnips are enjoyed in many regions of the world, but the French make a special celebration of the first young white turnips appearing the shops around Easter, when the sweet, peppery bulbous root can be enjoyed in one dish and the tender green top leaves in another.

WALNUTS
See Nuts and Seeds

WATERCRESS

Good source of:
- Beta-carotene, vitamin C, iron

Health benefits:
- Research suggests watercress contains a natural antibiotic, and it has been used to relieve stomach, urinary and respiratory complaints, and to heal pimples and other skin conditions.
- A member of the *Cruciferae* family, watercress contains substances that are thought to help reduce the risk of cancer.

Facts and tips:
- As far back as the twelfth century, watercress was praised by poets and healers alike; the latter using it as a tonic and means of preventing scurvy. Today it is a profitable farmed crop, used particularly in the spring as an ingredient in salad.

  It is hard to say whether this is a leaf vegetable or a herb, because it is used on its own – in soups and salads – and as a flavouring in butter and sauces. Watercress has a distinctive peppery flavour and contains health-giving natural antioxidants.

WATERMELON

Good source of:
- Safe water (about 95 per cent of flesh in a watermelon is water), vitamin C

Health benefits:
- Low-calorie dessert with the benefits of natural vitamins and minerals.

Facts and tips:
- When a ripe watermelon is tapped it sounds hollow. Do a little 'drumming' before you choose.

WHEAT
See Grains

WILD RICE
See Grains

WINE
See Miscellaneous Foods

YAMS

Sometimes confused with sweet potatoes, a vegetable with similar properties.

Good source of:
- Beta-carotene – the darker the yam, the higher the content

Health benefits:
- Studies in America suggest as little as half a cup a day of yams or other dark orange vegetable may reduce the risk of lung cancer by 50 per cent.
- Oestrogen-like substances in yams can help control menopause symptoms.

Facts and tips:
- These root vegetables originated in Africa and were transported to North America and the Caribbean as an easy to

grow and familiar food for transported slaves. They reached Europe when people from the West Indies crossed the Atlantic to find work in the growing industrial communities.

YEAST EXTRACTS

Good source of:
- Folate, B vitamins (most contain vitamin B_{12}), potassium, magnesium and zinc. The yeast in some products is cultured in media containing chromium in order to enrich the extract with this mineral.

Heath benefits:
- Excellent source of vitamins and minerals.

Facts and tips:
- People susceptible to *Candida albicans* infections may find eating yeast extract makes their condition worse.

People treated with antidepressants containing MAOIs (monoamine oxidase inhibitors) should avoid using yeast extracts because they may interact with the drug and cause a dangerous rise in blood pressure. Ask your doctor for advice.

ZUCCHINI
See Courgette

GRAINS

Five out of every 10 food choices we make should originate from grains: that makes sandwiches, pasta, pizza, noodles, couscous, popcorn, oatmeal and other breakfast cereals, chapattis and a host of other good foods essential for a healthy diet. Rich in slow-release carbohydrate that maintains a healthy blood sugar level over a long time, grains also contain the proteins, minerals and vitamins we need.

Wholegrains and wholegrain flour are excellent foods because they are rich sources of B-complex vitamins, vitamin E, fibre, iron, essential fatty acids and proteins. Be aware that not all proteins in grains contain the full mix of amino acids we need; wild rice is an (expensive) exception. By mixing grains with other protein foods this is altered, however *(see also pp.42–4 and 222)*.

Different parts of the world – and therefore different cuisines – rely on different grains for cooking.

Grains and world regions

Japan	rice
China	rice, millet and wheat
India	rice, wheat, sorghum
Central and South America	corn, amaranth, Quinoa (a staple grain of the Incas, now popular in the United States)
Africa	sorghum (millet)
Middle East	wheat, barley
Russia (also Eastern Europe)	buckwheat
Northern Europe	oats, wheat, rye, barley
North America	corn, wheat and rice

Modern tastes favour 'polished' grains, which have had not only the outer husks but also the rough surface of the grain kernels removed; this removes many key nutrients, and smart eaters choose either wholegrain products (which have had only the indigestible outer husk removed) or products which have been fortified with vitamins and minerals. White bread, by law, must be fortified with thiamin, niacin, calcium and iron to make up for some of the nutrients lost during the refining process.

BARLEY (WHOLEGRAIN)

Good source of:
- Complex carbohydrate, tryptophan, soluble fibre, vitamin E, folate, most B vitamins, iodine, potassium, calcium, magnesium, phosphate, iron, copper, zinc, manganese, some unsaturated fatty acids, about 10 per cent protein

Health benefits:
- Soluble fibre helps lower blood cholesterol levels.
- Barley has a low gluten content. As its flour makes excellent baked goods, this is a great boon to those who are sensitive to gluten.
- Wholegrain barley is a good source of most nutrients.

Facts and tips:
- Thought by archaeologists to be the earliest grain cultivated, barley is used in various cuisines as wholegrain (husk removed), polished grain, flakes and flour. It is an excellent thickening in soups and stews, and is favoured as an ingredient in certain brewed drinks.

  Barley adds weight and taste to vegetable soups that may otherwise be slightly watery.

  'Pearl' barley, often used in cooking, has been stripped of its husk and partially milled to achieve its round form. Many nutrients are lost in the process.

BRAN (WHEAT)

Good source of:
- Fibre, folate, vitamin E, biotin, most B vitamins, most minerals (excluding iodine)

Health benefits:
- Helps maintain a healthy bowel by speeding food through the digestive tract. This prevents carcinogenic substances remaining in the body long enough to have an effect.

Facts and tips:
- An analysis of the outer husk of wheat bran has shown it to be surprisingly high in vitamins and minerals. (*See Cereals and Cereal Products*, The Royal Society of Chemistry, Ministry of Agriculture, Fisheries and Foods, 1988.)

  During the 1980s and early 90s bran was in favour with nutritionists and medical scientists alike. Studies of illness in native populations showed a link between low levels of gastric cancer, especially cancers of the bowel, and high levels of fibre in their diet. While it is still accepted that bran and other forms of plant fibre reduce constipation and other bowel complaints, including piles, it is now also understood that too much uncooked bran in the diet can have negative effects.

Raw bran contains phytic acid, a natural compound that blocks the body's ability to absorb calcium, zinc, iron and magnesium. Cooking (baking and toasting, for example) breaks down phytic acids and removes this danger. (Note: Phytic acid may also act to block some cancer-causing substances.)

There are two types of fibre in bran: soluble and insoluble. Oat bran is rich in a soluble fibre which has been shown to lower blood cholesterol levels when added to the diet in fairly high levels. It is thought that this fibre binds the cholesterol in food and causes it to be expelled from the body rather than absorbed into the bloodstream.

BREAD, PASTA AND OTHER FOODS MADE WITH FLOUR

Good source of:
- Complete carbohydrate, fibre, protein, some fats, vitamins and minerals

Health benefits:
- *This is the most important source of nutrition in the human diet. About half of our calories should come from foods made with wholegrains and wholegrain flour.*

Facts and tips:
- Mixtures of ground grains (wheat, corn, rice and others) mixed with water and baked on a flat surface or in a simple oven are some of the oldest foods in the world.

The slow-release carbohydrate, the balance of minerals and vitamins, and the filling sensation these foods bring to our meals make them our most valuable single source of nutrition. Do not choose the modern milled and bleached varieties, however, rather enjoy the dark and flavourful products that still contain the rough and nutrient-rich parts of grains that milling removes. No added supplements can ever restore, in content or balance, the blessing that Nature bestows through these foods.

Remember:
- Most grains lack one or more of the essential amino acids needed for human nutrition. Mix grains with other foods, including beans, to obtain the full mix of protein building blocks your body needs.

BUCKWHEAT

Good source of:
- Fibre, the essential amino acid lysine, vitamin A, complex carbohydrate, protein, calcium, selenium

Health benefits:
- May be useful in controlling sugar metabolism and therefore useful in the diet of diabetics.
- Ideal for people allergic to wheat.
- Fibre helps control constipation.

Facts and tips:
- An ancient grain originating in the Orient and brought to Europe during the fourteenth century, this is not a wheat, but the seed of a fruit related to rhubarb. When cracked and toasted the grain is known as *kasha* and is a popular food in Eastern Europe.

  Slightly lower in calories that other grains, buckwheat does not make good bread, but makes an excellent flour for use in griddle-baked flat bread (*galettes*) and *crêpes*; the *crêperies* in Brittany are well known for their delicious savoury buckwheat *crêpes*, stuffed with mushrooms or a whole egg.

BULGHAR (BULGUR)

Good source of:
- Potassium, B vitamins, fibre, iron and calcium

Health benefits:
- Fibre helps speed food through the digestive tract, thus reducing the effects of cancer-causing chemicals.

Facts and tips:
- Bulghar is wheat which has been partially cooked, cracked and dried. It is popular in Greece and around the Mediterranean. Nutty and sweet, cracked bulghar is delicious with roasted vegetables and served as *Tabbouleh*, a salad of steamed cracked wheat, chopped tomatoes, onions and mint, and dressed with lemon juice and olive oil. The nutritional value is about the same as other types of wheat.

CORN (MAIZE, SWEETCORN)

Good source of:
- Natural energy, protein, dietary fibre, polyunsaturated oil, vitamin A and potassium

Health benefits:
- Helps maintain natural bowel function.
- The essential fatty acid in whole corn helps build and sustain healthy body tissues and fights the development of degenerative illnesses.

Facts and tips:
- A native of North America, this is one of the most versatile grains and finds its way to tables throughout the Americas and Europe in many forms, including corn on the cob, kernels, popcorn, corn starch or corn flour, grits, corn oil, corn syrup (used widely in the manufacture of sweets or candies) and cornmeal. Cornmeal, a favourite primary ingredient in America for making quick breads and batter cakes, is also used in many parts of Europe, in a slightly finer form, to make *polenta*, a rather thick savoury pudding.

  Milling corn strips it of many nutrients; it is better to use whole kernels or ground cornmeal. The protein in corn is gluten-free.

  The French still find corn a 'coarse' vegetable, and primarily use it for feeding poultry and livestock.

  See also Oils, pages 165–9.

MAIZE
See Corn

MILLET (SORGHUM)

Good source of:
- Complex carbohydrate

Health benefits:
- Gluten free, but high in the proteins, vitamins and minerals needed for good health.

Facts and tips:
- Millet 'bulks up' with water when cooked and makes a filling and relatively low-calorie accompaniment for spicy vegetable stews. Added to soups, it adds thickness and texture.
- An ancient and staple food grown throughout North Africa, southern Europe and Asia.

OATS

Good source of:
- Essential fatty acids (both omega-3 and omega-6), carbohydrates, protein, potassium, iron, B-complex vitamins and soluble fibre

Health benefits:
- When oats are cooked, a thick jelly-like substance is released. This contains vegetable gums, pectin and soluble fibre, which are thought to help control blood cholesterol levels. Oats are a particularly rich source of soluble fibre.

Facts and tips:
- Native to the damp, chilly terrain of northern Europe, Ireland and Scotland, oats have been a staple food since ancient times. Today they appear on our supermarket shelves as whole oats, oat flakes (rolled oats), oatmeal and oat bran (the hard outer shell of the oat kernel).

  Oat bran became a darling of the diet industry in the early 1960s, when medical scientists suggested it helped control blood cholesterol levels. Studies in the United States suggest that people with clinically elevated blood cholesterol may be helped by adding 5 grams (0.17 oz) of oat bran to their diet each day. Why this happens is not understood, but chemicals in oat bran fibre may reduce the production of cholesterol, absorb cholesterol-based bile acids in the gut and help transport them out of the body.

RICE

Good source of:
- Carbohydrates, B-complex vitamins, calcium and phosphorus

Also contains:
- Protein, but not as much as some other grains, especially varieties of hard wheat

Health benefits:
- The healing powers of rice have long been recognized in treating problems of the digestive system, including diarrhoea, constipation, indigestion and diverticulitis. Rice water (make by boiling rice until it disintegrates and then straining it) is an excellent source of nutrition for sick children and elderly people who have trouble eating or swallowing.

  Rice bran is thought to reduce the risk of bowel cancer.

Facts and tips:
- The principal source of nutrition for half the people in the world, rice has been cultivated for over 3,000 years. Food historians think it first spread from India to China, and then on to the Korean peninsula and the Philippines, before going on to Indonesia and Japan. When Alexander the Great invaded India in 327 BC he brought rice back to Greece, from where its cultivation spread to North Africa, Europe and many colonies of European powers. Today, rice is a major cash crop grown in places as dissimilar as China, the United States and the marshy Camargue, southern France.

  Popular types of rice include Aboria (used in Italian risotto), Basmati (sweetly scented rice favoured by cooks in

India), jasmine (used in Chinese cooking, having a slightly floral scent) and Carolina (a long-grained variety grown in the United States, known for its fluffy and dry appearance after cooking).

Rice is digested slowly, thus releasing sugar into the blood over a long period of time. This is ideal for diabetics. It is gluten-free.

White rice (polished rice) has had its outer layer removed, and with it most of its protein and vitamins. Some types of white rice are fortified with thiamin; when you buy this food, read the packaging to see which nutrients you are getting.

Caution:
- Rice contains phytic acid, which inhibits the absorption of calcium and iron during digestion. Avoid macrobiotic diets based on brown rice; they can contribute to calcium and iron deficiencies which may be particularly bad for children.

RYE

Good source of:
- Iron, phosphorus and B-complex vitamins

Health benefits:
- Excellent source of basic nutrition.

Facts and tips:
- Native to western Asia, rye became a staple grain in Eastern European countries after its introduction there. Although it does not contain enough gluten to produce a good bread by itself, it is often used in combination with wheat to make the dark wholemeal breads favoured in many forms of European ethnic cooking.

  Rye is also used to produce a variety of alcoholic beverages, including whisky, vodka and gin.

SORGHUM
See Millet

SWEETCORN
See Corn

WHEAT

Good source of:
- Protein and complex carbohydrate for quick energy, minerals (iron, phosphorus), vitamins (especially vitamin E) and essential fatty acids. Most of the oils and fat-soluble vitamins are contained in the germ – or embryonic – portion of the wheat grain, or seed. The much larger core of the grain contains the protein and carbohydrate and most of the vitamins. Wheatgerm oil is an important source of nutrition.

Facts and tips:
- A grain known to have been cultivated since the Neolithic Age, wheat now forms the basis for a majority of foods in the Western world. Through genetic selection, a wide range of wheats are now produced that differ in their gluten content, kernel hardness and size, and requirements for maximum productive growth.

  Sprouted wheat is a recent addition to our supermarket shelves. You can produce your own by placing some grain in a flat container, covering it with water (I prefer to use filtered water) and leaving it in a warm place for 24 hours. Rinse, cover with water and leave for another 24 hours; finally, rinse and use. The sprouts will be white and fairly short. This process increases the protein and vitamin content of the grain, and provides good flavour and texture when used in a salad or combined with other ingredients in home-baked breads.

Warning:
- Some people are allergic to gluten and should avoid wheat and all products containing wheat flour. Good alternatives are buckwheat and buckwheat flour, cornflour and cornmeal, rice and rice flour, lentils, millet and wild rice.

WILD RICE

This is not a rice, but the seeds of a wild aquatic grass found in Canada and the northern parts of the United States. It is rich in B-complex vitamins and high in protein (14 per cent by dry weight). *Wild rice contains complete protein, including all essential amino acids*, and is particularly rich in lysine. This is an expensive grain because it is harvested by hand; however, it swells to four times its bulk during cooking, and mixed with Basmati rice, it becomes a delicious part of a well balanced diet. Adding nuts and raisins to wild rice complements its flavour and texture, and adds nutrients. Cold cooked wild rice can form the basis of a fine and healthy salad – just toss with cold vegetables and dress with oil (olive, walnut or corn oil) and fresh lemon juice or vinegar.

Wild rice needs to be cooked for a much longer period than regular rice – from 30 to 60 minutes when cooked from dry – however, this time can be reduced by soaking the grain in filtered water overnight.

HERBS

Chemists, botanists and doctors are the direct descendants of the old alchemists, herbalists and magicians.

Dr Conrad Gorinsky

Although this is a book about cooking with plants and treats herbs only as ingredients in what we eat, not for the specific purposes of healing, they must be acknowledged for both their culinary and healing qualities. Records of plants grown together to nurture and to heal stretch back to the earliest civilizations and extend through to modern times. From Egyptian hieroglyphs of harvests and healing, through the teachings of Hippocrates and beyond, stories of food, plants and health have been intertwined. Echoes of past times remain. Medieval monks grew herbs and vegetables within the walls of the monasteries; was this the origin of our traditional kitchen garden, with its fresh radishes and salad greens accompanied by parsley, mint and chives? While in India, a physician's skills were not limited to medical matters but to the kitchen as well, where he advised on diet and the blending of foods and flavours to accommodate the season and the health of his master. Today Indian food lacks character without its powerful blends of energy-rich, sustaining plants, like rice and lentils, and flavourful herbs, such as coriander, cardamon and cloves, which add aroma, taste and healing goodness.

Much of the mystery of cooking and enjoying food rests in the blends of flavours and textures. We miss much of this excitement in our modern diet because so many foods are processed and pre-packaged. Until we prepare more of our food at home, from simple, fresh ingredients, we will continue to miss some of the pleasure we can receive from preparing truly healthy food. What would tomato and cheese pizza or a plate of pasta dressed with tomatoes and olive oil be without the complementary character of basil? Would curry be the same without fresh coriander? And how many great sauces and egg dishes would be altogether different without the presence of parsley?

Nutrients in herbs

Herbs are surprising – and often overlooked – sources of nutrition. As most are bright green leaves, they are loaded with chlorophyll and all of the minerals and vitamins required for photosynthesis activity.

Carotenoids and other plant pigments and phytochemicals are there to aid and strengthen the human body. Some phytochemicals are well known to the chemist – thymol, for example – while others are only now being systematically explored for their health benefits. Expert herbalists have known about these benefits for years; the subject of essential oils from plants is an exciting topic in alternative health care, for instance. Unfortunately, it has taken traditional medical scientists much longer to accept that essential oils and other powerful substances in these flavourful plants have remarkable healing properties.

One herb is so packed with nutrients, so complete in its nutrient value, that it must be given special mention here: parsley. This is a surprisingly beneficial plant, even when used as a herb. Because it grows quickly – and almost anywhere there is sunlight and a little water – most cooks can enjoy picking and using fresh sprigs from their own plants. For the cost of a few seeds and minimal trouble, chopped parsley will add the following to the salads, sauces, soups and main dishes you prepare: vitamin E, vitamin C, folate, all of the B-complex vitamins, iron, potassium, calcium and a range of other minerals. In addition, it is a rich source of carotene. The benefits of beta-carotene, and probably other nutrients, appear to work best when they are consumed as part of a naturally occurring balance of nutrients. That makes parsley an important part of the fight against cancer and other degenerative diseases of Western culture.

BASIL

Good source of:
- Carotene, thiamin, niacin, iron

Health benefits:
- Good for calming the nerves and soothing stomach cramps.
- Also helps relieve nausea.

Facts and tips:
- Basil is a soft herb at the heart of the taste and smell of the Mediterranean diet. Stand in a sun-drenched garden well planted with basil, breathe in deeply and enjoy. You will experience one of the most delightful and complex fragrances in the plant world. There is no comparison between the aroma and flavour of the herb in its fresh and dried form. The tender green leaves of fresh basil are rich in delicate essential oils with subtle flavours which are entirely lost during the drying process. To fully enjoy basil, find a sunny spot in the garden and grow your own; or, if space and the climate do not favour that ideal arrangement, buy plants fresh from a market.

  Do not pick the leaves until just before use. Gently brush them free of any debris, or rinse them, and pat dry gently with a paper towel. Just before serving the stew, pasta sauce or salad in which you plan to use this princess of herbs, gently pull the leaves from stems, tear the leaves into large pieces and add to your dish. As the cook, and so the first to appreciate the escaping flood of essential oils, you will enjoy the full balance of fragrances captured in this plant.

BAY (BAY LAUREL)

Good sources of:
- Manganese, phosphate, iron and calcium

Health benefits:
- Manganese is important for normal bone development and forms a part of many enzymes.

Facts and tips:
- The bay laurel, or true laurel, is native to the Mediterranean region where, during ancient times, it was used to form crowns with which to honour heroes, poets and political leaders.

  A popular herb known to aid digestion, bay is a necessary part of a *bouquet garni* (a small tied bundle of herbs containing at least the following: a sprig of thyme, two or three bay leaves and one or two sprigs of flat parsley), while bay leaves add flavour to stews, soups and casseroles.

Warning:
- Do not cook with laurel other than from bay trees, as some types contain poisonous prussic acid.

BORAGE

Good source of:
- Minerals and some B vitamins; phytochemicals that may have medical significance

Health benefits:
- Used to fight respiratory infection and reduce pain from rheumatoid arthritis.
- This herb's name comes from the Arabic word *abūǎraq*, meaning 'sweat'; in fact, herbal tinctures of this plant are used to induce sweat.

THE PLANTS WE NEED TO EAT

Facts and tips:
- A perennial plant bearing edible blue flowers which add both flavour and colour to salads, soups and pasta dishes. Borage flowers are sometimes candied and used to decorate cakes and sweets.

CHERVIL

Good source of:
- Most minerals, some vitamin B_6

Health benefits:
- Contains phytochemicals that may have significance in controlling disease.
- Stimulates the digestion.

Facts and tips:
- A plant originating in Asia, but now known throughout the world for its unique aroma, chervil is as much at home in an elegant *sauce béarnise* as it is chopped and sprinkled over freshly grilled fish. One word of warning – the aroma and flavour of chervil are volatile, so do not add this herb until just before food is served. Excellent in egg dishes and sauces.

CHIVES

Good source of:
- Selenium (depends on soil), carotene, vitamin E, vitamin C, pantothenate, some B vitamins and some tryptophan

Health benefits:
- Selenium helps antioxidants work. Not all plants require this mineral in their own metabolism, so do not concentrate it in their tissues. Chives do use it and, depending on the amount of the mineral in the soil in which they are grown, concentrate it for our consumption.

Facts and tips:
- A member of the onion (*Allium*) family that can be enjoyed even when grown at home in clay pots on the terrace. The natural oils in chives have both anti-insect and anti-fungal qualities.

Chives stimulate the appetite and add interest to food. When cut in fine pieces, they are good with salads or sauces served to the picky eater or convalescent.

CORIANDER

Good source of:
- Vitamin B$_6$, folate

Health benefits:
- Herbalists use coriander leaves and seeds to help strengthen the urinary system and act as a tonic for the stomach and heart.
- Thought to be used in many chillies and spicy dishes as an aid to digestion and a natural antibiotic.

Facts and tips:
- A plant prized for both its leaves and seeds, coriander was used by both the ancient Hebrews and Romans. For the Hebrews, it was an excellent flavouring for bread; the Romans found it an ideal flavour for preserved meat. Experts claim Charlemagne liked this herb so much he personally encouraged its cultivation. It was used by the ancient Carthusian monks in the formulation of the herbal liqueur we now enjoy as Chartreuse.
- Now used widely in Asian cooking, the tender green leaves have many names, including 'Greek parsley', 'Arab parsley' and 'Chinese parsley'.

CRESS

Good source of:
- Vitamin C, carotene, folate, calcium and iron

Health benefits:
- Folate is important for normal development of the human brain and spinal cord.

Facts and tips:
- A member of the mustard family, cress is rich with phytochemicals that help fight cancer.

DILL

Good source of:
- Carotene, folate, vitamin C, thiamin, riboflavin, iron, manganese, zinc, potassium and calcium

Health benefits:
- Helps flatulence and gripe in children.
- The Romans used dill as a symbol of vitality.
- Dill seeds are an excellent digestive.

Facts and tips:
- With a distinctive flavour mildly resembling aniseed, or liquorice, either the feathery leaves or seeds of this garden annual are added to food. It goes well with egg dishes, in vegetable cream soups, cottage cheese and raw cucumber salad.

  Dill is another native of Asia, brought to the West along the ancient trade routes.

FENNEL (SEEDS)

Good source of:
- Potassium, calcium, manganese, zinc and other minerals

Also contain:
- Some thiamin, riboflavin and niacin

Health benefits:
- The seeds and the feathery leaves of fennel are both used to control vomiting and nausea, and prevent wind. Chewing the seeds is said to help control hunger.
- Fennel is rich in phytochemicals that appear to affect the processes of the human body.
- Minerals are important parts of all body tissues, not just bone. The acid-base balance in blood and other body fluids depends on minerals consumed as part of food.
- The great herbalist Nicholas Culpeper advised nursing mothers to drink fennel tea to stimulate the flow of milk; a spoonful or two of the same tea would ease her child's colic.

Facts and tips:
- Fennel is a member of the *Umbelliferae* family and related to dill, coriander and chervil. It was probably first introduced into the United States by Thomas Appleton, the American Counsel to Leghorn, Italy, in 1824, when he sent seeds to the retired President Thomas Jefferson for his garden at Monticello.

MARJORAM

Our common sweet marjoram is warming and comfortable in cold diseases of the head, stomach, sinews, and other parts...
Nicholas Culpeper

Good source of:
- Potassium, calcium, iron, magnesium, manganese, zinc, carotene and niacin

Health benefits:
- Thought to be good for digestion and fighting symptoms of an oncoming attack of 'flu or a cold.
- Contains thymol, a powerful antiseptic.

Facts and tips:
- There are two types of this herb: sweet marjoram and wild marjoram, more often called oregano. Sweet marjoram has a rather mild flavour and is used widely in European salads, vegetable and meat dishes. Wild marjoram, or oregano, has a more powerful flavour and is needed to add much of the characteristic flavour to Mediterranean food and Italian-style dishes. Drying intensifies the flavour of both; use with some caution.

Warning:
- This is a powerful herb and should not be used for medical reasons during pregnancy.

THE PLANTS WE NEED TO EAT

Good source of:

- Carotene, vitamins C and E, folate and certain terpines which act as powerful natural antioxidants

Health benefits:

- Loaded with natural antioxidants, mint helps strengthen the immune system.
- The distinctive essential oil in mint is menthol, known to refresh the taste buds and sooth stomach discomfort. It also helps fight the early signs of a cold.

Facts and tips:

- A wonderfully versatile and fragrant plant that adds pleasure to drinks, desserts and a wide range of classic and ethnic foods. There are about two dozen varieties of mint, each with its own individuality; however, all can be a curse if they are allowed to run free in your garden. Plant in clay pots before placing in the herb border, or they will take over! Some of the more popular varieties are spearmint, pineapple mint, lemon mint and peppermint.

  Mint is an excellent cooling ingredient in summer food and drink.

OREGANO
See Marjoram

PARSLEY

The nutrient-packed star of the herbs!

Good source of:

- Vitamin C, beta-carotene and other carotenoids, folate and iron

Health benefits:

- Packed with beta-carotene, vitamin C and a host of important phytochemicals, parsley helps fight cancer and heart disease, encourages the healthy growth of infants and children, promotes healing and helps ward off infection by strengthening the immune system.

- Parsley is almost unique among herbs in the amount of vitamin C it contains.

Facts and tips:
- Once believed by the ancients to have magic powers, our respect for parsley has slumped to mild tolerance, its viability on our tables only secured by the ease with which it brightens up a platter of cream-coloured, often nondescript food. In fact, it still has magical powers – it is packed with nutritious vitamins and minerals, and beneficial phytochemicals, such as chlorophyll and beta-carotene.

  Whenever you see a restaurant plate swept clean by a diner, except for a lonely, crisp-looking sprig of parsley, see it as an opportunity missed; one of the best parts of the meal is going to end up in the rubbish. Even if people stare, eat your parsley garnish. It probably contains more micro-nutrients than the main dish and certainly more than the dessert.

  Parsley acts as a natural breath-freshener. Serve during the same meal as garlic.

ROSEMARY

Good source of:
- Carotenes, potassium, calcium and zinc

Health benefits:
- Contains phytochemicals, called quinones, that under laboratory conditions inhibit the action of certain groups of carcinogens.
- Thought to stimulate both the circulatory and nervous system. May help indigestion and flatulence.
- Tea made from rosemary is sometimes used to treat a nervous headache and as an antiseptic gargle. Rosemary contains natural antiseptics.

Facts and tips:
- A rapidly growing shrub originating in the Mediterranean region and imparting a distinctive flavour that lifts the taste of bread, meats, salads and even the lowly potato. The needle-like leaves of this plant are packed with its distinctive

essence and only a few are needed to add flavour to a marinade or grill. Although this herb is often associated with the cooking of meat, especially lamb and pork, it also goes well with tomato dishes and roasted vegetables dressed in olive oil. The flowers can be used to add both beauty and aroma to a salad.

SAGE

Jupiter claims this, and bids me tell you, it is good for the liver, and to breed blood.

Nicholas Culpeper

Good source of:
- Carotenes, some thiamin, calcium, iron, magnesium, manganese and zinc

Health benefits:
- Dried sage is a good source of a number of important minerals in natural balance. Zinc and magnesium are both important to maintaining healthy bones and may be vital in controlling appetite.

Facts and tips:
- Aids the digestion of rich foods. Sage tea can be used as a gargle for a sore throat and to help control excessive sweating. It is also sometimes used to calm anxiety. Used by the French to fight fatigue and stimulate the nervous system.

   A perennial herb that grows well in temperate to warm climates, where the warm sun concentrates the oils responsible for its special flavour, sage is indispensable for the flavours of many regional dishes – rice minestrone and *piccata* in Italy, for example, and the stuffing for poultry and pork in England. It also adds excellent flavour when shredded and added to vegetable soups and stews. Marries well with rich foods.

SAVORY

Good source of:
- Carotenes and minerals; some B vitamins

Health benefits:
- Has a mild antiseptic quality.
- Helps maintain a healthy digestive system and controls flatulence.
- Often used with beans and lentils to aid their digestion.

Facts and tips:
- Believed by.the Romans to have aphrodisiac qualities, savory is used today to flavour soups, pulses, Provençal salads and various pâtés.

TARRAGON

Good source of:
- Carotene (when dried)

Health benefits:
- Helps control stomach acidity and flatulence. An excellent digestive and general tonic.

Facts and tips:
- Once erroneously believed to cure snakebite, this ancient herb originated in Central Asia. Delicious raw in salads and mixed with cooked beans, it adds a distinctive flavour and aroma to pickles. Fresh tarragon is always best.

THYME

Good source of:
- Calcium and carotene

Also contains:
- Other minerals and B vitamins

Health benefits:
- Thyme is one of the great and long-respected healing herbs. According to the great English physician Nicholas Culpeper, 'It is a noble strengthen of the lungs, as notable a one as grows... It purges the body of phlegm...' and 'The herb taken any way inwardly, comforts the stomach much, and expels wind.'

- Contains an essential oil – thymol – which is a natural antiseptic.
- Other natural chemicals in thyme aid digestion and are claimed to help cure a hangover.

Facts and tips:
- A perennial plant favoured by gardeners because of its many colourful varieties, thyme is a basic cooking ingredient. It adds character to pulses, egg dishes, pasta, tomato dishes and stews. It may be used to flavour certain alcoholic beverages and is a required part of a *bouquet garni*.

  Thyme is a good natural antiseptic, and a tea made from fresh leaves can be used as a gargle and to help control catarrh and other symptoms of 'flu or a cold. This aromatic herb also gives a boost to the nervous system.

NUTS AND SEEDS

Nuts and seeds add crunch and flavour to foods, and are a rich source of essential fats, minerals and vitamins, proteins and fibre. Unfortunately, their high calorie content causes many people to shy away from them. When making up your mind about where these foods fit into your diet, remember that they are packed with goodness and a little goes a long way. Nut lovers may take comfort from the fact that recent studies suggest they make a significant nutritional contribution to what we have come to know as the Mediterranean diet.

In general, nuts are rich in vitamin E and the B vitamins, especially thiamin and niacin. Some are excellent sources of folate and biotin. But remember, like other food sources from plants, they lack vitamin B_{12}.

Nuts are also a rich source of protein and essential fatty acids; several – walnuts for example – contain both omega-3 and omega-6 fatty acids. However, like other plant protein sources, they do not contain all of the essential amino acids, although some contain significant quantities of the amino acid tryptophan. To meet your protein requirements, enjoy them with other foods: peanut butter on wholemeal bread, hazelnut butter with bean sprouts on pitta bread, for example. Be creative! How about taking advantage of the amino acids in sprouts and combining peanut butter with alfalfa sprouts in a wholemeal bap (bun)?

Warning: A small percentage of the population is allergic to nuts, particularly peanuts (groundnuts) and the results can be fatal. If you have any indication that you or a member of your family suffers from a nut allergy, make sure you read all food packaging carefully to make certain no nuts or products made from nuts are used as ingredients. Nuts do not have to be obvious in a product to cause an allergic reaction.

A mould that grows on peanuts has been shown to produce a powerful carcinogenic toxin that attacks the liver. Buy nuts and nut products from reputable sources, and if you use unroasted peanuts at home, make certain they remain dry and free of any mouldy contamination.

Here are the nutritional values of the main nuts and seeds:

Almonds – A good source of protein and monounsaturated fats, calcium, magnesium, selenium, iodine, vitamin E, folate and most B vitamins. Almonds may help control blood cholesterol levels.

Brazil nuts – An excellent source of selenium and iodine; high in zinc, magnesium and other minerals. Good source of vitamin E, contain B vitamins and folate. Contain about equal quantities of saturated, mono- and polyunsaturated fats.

Cashew nuts – A good source of folate, biotin and pantothenate. Contain useful quantities of selenium and iodine. Rich in monounsaturated fatty acids.

Hazelnuts – A rich source of monounsaturated fats, like those found in olive oil. Good source of iodine, manganese, zinc, folate, biotin, tryptophan and protein. A very good source of vitamin E.

Peanuts (groundnuts) – A good source of both mono- and polyunsaturated fats. Very good source of vitamin E, biotin, folate, pantothenate, niacin, thiamin, iodine, selenium, zinc and other minerals. Good source of protein.

Pecans – A rich source of monounsaturated fatty acids. Contain some carotene and most B vitamins. Good source of tryptophan.

Pine nuts – An excellent source of polyunsaturated fats. Good source of iron, copper, zinc, magnesium, manganese and potassium. Excellent source of vitamin E, contain tryptophan and some B vitamins.

Pumpkin seeds – Contain zinc, magnesium, phosphorous and potassium, also useful essential fatty acids and vitamins.

Sesame seeds – A good source of both mono- and polyunsaturated, zinc, iron, phosphate, magnesium, calcium, copper, biotin, pantothenate, folate, vitamin B_6, most other B vitamins and vitamin E.

Sunflower seeds – A good source of polyunsaturated fats, iron, phosphate, magnesium, manganese, carotene, tryptophan and some B vitamins. Excellent source of vitamin E.

Walnuts – An excellent source of polyunsaturated fats, including both omega-3 and omega-6 types, biotin, selenium and iodine. Good source of other essential minerals, vitamin E, folate and most other B vitamins.

OILS

Oils are important foods. They are needed for frying and grilling, add texture to foods, aid the combination of ingredients, and supply flavour and 'mouth appeal' to what we eat. If anyone doubts the important of edible oils, all they need do is look at the multi-billion dollar a year business that extracts, ships and supplies seed and nut oils to outlets around the world.

Oils and fats were partially covered in Chapter 2 *(see pp. 44–5)*; now we will look at what they contain.

Oils are about 99 per cent fat. That can be polyunsaturated fat, which is high in essential fatty acids, monounsaturated fat (mainly a fat called oleic acid) and saturated fat. A few nut oils – coconut oil is the best case – are surprisingly rich sources of saturated fats. But most oils consist of a blend of the healthy unsaturated varieties.

Saturated fats in foods – from animal products and processed foods made with animal fats – are linked by medical research with serious illness. 'Furring' of the arteries in coronary heart disease, increased blood cholesterol levels and greater risks from breast and colon cancer have all been seen to accompany eating diets too rich in saturated fat. This is a health problem that has an easy answer: shift your food choices away from animal products and towards foods from plants. Not only will you eat less dangerous saturated fat and more essential polyunsaturated fat, but you will also be eating a lower ratio of fat in your total diet. According to modern medical thinking, your daily fat intake should represent no more than 30 per cent of the total number of calories eaten. With the exception of seeds and nuts – which are packed with energy and goodness to start off the next generation of plants – most plant tissues consist primarily of starchy carbohydrates, fibre and protein.

Oils also contain some fat-soluble vitamins, a few minerals and some very valuable fat-soluble phytochemicals that contain the flavour and aromatic parts of the oil. Vitamin E is a key ingredient in good oils.

However, as any cook will agree, not all oils are the same. Some carry the natural aroma and taste of the nut or seed from which were produced, while others have little character at all. The reason for these differences rests in the methods used to extract and purify oils. The more an oil is purified, the more it loses its nutrients and individual character. Some of these processes involve the heat extraction of oil from tons of fibrous seed or nut pulp; other methods involve using the same pulp but extracting the oil by chemical means. As a general rule, the harsher the procedure, the more likely the oil will lose some of its goodness.

The classic example of how extraction affects the final product is olive oil. Cold-pressed, extra virgin olive oil has had the least done to it. When you look at a bottle you will see it has a deep colour and sometimes small bits floating at the bottom. This is rich with plant chemicals that promise flavour, aroma and good nutrition. Increasing the processing produces virgin oil which is lighter in colour and has less 'umph'. And finally, when the presses have worked overtime and all of the extraction processes have wrung the last of the oil from the olive pulp, a nice golden oil remains. It is cheaper than the others and has fewer nutrients – but remember that it too has a place on your kitchen shelf. For the valuable nutrients in virgin, or unrefined, oils reduce an oil's capacity to resist heat and cause it to 'smoke', or chemically break down at a fairly low temperature. By removing these substances, some of the character you want on a fresh salad will be removed, but what remains is far better for frying and cooking with high temperatures.

Pure oil is made up of fat, almost all of which consists of molecules called triglycerides. These are nothing more than a molecule of a substance called glycerol with three long carbon chains, called fatty acids, attached to it. These fatty acids tell the real story about fat and health. If they are a simple chain with no chemically weak points (unsaturated double-bonds), they are 'saturated'; if there is only one of these weak points in the chain, they are mono- (meaning 'one') unsaturated; if there are two, three or more of these double-bonds, they are poly- (many) unsaturated.

The human body cannot function without a steady supply of very specific, essential, unsaturated fatty acids. Plant oils

---

Table V: The types and amounts of fats in 100 grams (3.5 oz) of various seed and nut oils

Plant oil	Saturated fat/grams	Monoun- saturated fat/grams	Polyun- saturated fat/grams
Coconut oil	85.2	6.6	1.7
Corn oil	12.7	24.7	57.8
Olive oil	14.0	69.7	11.2
Rapeseed (canola) oil	6.6	57.2	31.5
Safflower oil	10.2	12.6	72.1
Sunflower oil	11.9	20.2	63.0

From McCance and Widdowson's *The Composition of Foods*, fifth edition (The Royal Society of Chemistry and the Ministry and Agriculture, Fisheries and Food, 1993).

Note: The higher the polyunsaturated content of an oil, the more quickly it is damaged by heat. This damage destroys its nutritional goodness and can create harmful substances. In the above examples, olive and rapeseed oils are good for frying. Sunflower, safflower and corn oil are best used in dressing and sauces. Coconut oil should be considered a saturated fat.

are our best sources of these. As already mentioned, there are two types of essential fatty acids, known as omega-3 and omega-6. Research shows we need more omega-6 than omega-3 fats (often associated with fish oil). If you want more specific information on these, read my book *The Fats We Need to Eat* (Thorsons, 1995). For now, just remember to explore and use more oils from plants in your diet. Over time your body will benefit and your sensual experience with food will profit from the exciting variation nut and seed oils bring.

SPICES

Spices, like herbs, add nutritional value to foods; most often these are minerals and some of the B vitamins. However, they are also rich sources of phytochemicals and associate nutrients that can influence our health. When you add spices to your cooking, you not only add aroma and flavour, you also add Nature's miracle nutrients.

BLACK PEPPER

Good source of:
- Potassium, calcium, magnesium, sulphur, manganese, riboflavin and tryptophan

Health benefits:
- The blend of minerals and volatile oils in pepper increases the appetite and is said to aid digestion.

Facts and tips:
- Treasured for its aroma and fiery taste in Asia before it was introduced into Greece by Alexander the Great, pepper is one of the most important universal spices. Its value is decreased only by modern contempt for the available. The hot, oily aroma of freshly ground black peppercorns adds gusto to almost every type and ethnic variety of food, from the aromatic French *steak au poivre* and German *Pfefferkuchen* (pepper cake) to the American sweet cucumber pickles.

CARAWAY

Good source of:
- Most essential minerals, including manganese, iron and zinc

Also contains:
- Thiamin, riboflavin and niacin

Health benefits:
- It is thought to aid digestion and sweeten the breath.

Facts and tips:
- Caraway is a plant valued since prehistory for its elongated seeds, which taste mildly of liquorice. They are an international ingredient. Shakespeare mentions them in *Falstaff*, while in Eastern Europe they add distinctive character to stews and cabbage dishes, and many cultures enjoy them when they are blended with the ingredients in alcoholic beverages.

CINNAMON

Good source of:
- Calcium, iron, magnesium

Health benefits:
- Some research suggests phytochemicals in cinnamon may aid the metabolism of glucose, thus helping control blood sugar levels.

Facts and tips:
- This ancient spice is mentioned in Sanskrit and biblical texts. It consists of the bark of tropical cinnamon trees, which is removed, rolled into a cylinder and dried before sale.

CUMIN SEEDS

Good source of:
- Potassium, calcium, magnesium, phosphorous, iron, copper, zinc and manganese

Also contain:
- Some thiamin, niacin and riboflavin

Health benefits:
- Aid digestion.

Facts and tips:
- These spindle-shaped seeds with a hot, slightly bitter taste are cultivated from northern Europe to the Mediterranean. Used since ancient times to preserve meat and flavour fish dishes and sauces, today they are used most often as a distinctive flavouring in Munster cheese and regional breads, especially those made in Eastern Europe.

CURRY

Good source of:
- Antioxidants and phytochemicals that act as associate nutrients; a curry is a powerhouse of goodness

Health benefits:
- Antioxidants may help control certain forms of cancer and the degenerative diseases associated with Western society.
- Some of the associate nutrients in curry have been shown to lower cholesterol levels, thus reducing risks from stroke and heart disease.
- The antioxidants and peppery chemicals in curry may reduce the pain of arthritis.

Facts and tips:
- Curry is a blend of spices, although it is sold as a single product.

  Common ingredients are cardamon, chilli powder, cinnamon, coriander seeds, cumin, fenugreek, garlic and ginger, mustard, tamarind and pepper. Sometimes coconut milk (high in saturated fat) is added for flavour and a smooth texture. The exact blend of spices differs between ethnic groups.

FENUGREEK
See Spices

Good source of:
- Iron (contains more iron than watercress by weight)

Health benefits:
- Stimulates digestion.

Facts and tips:
- The tiny black seeds of this herb are sprouted and used as a vegetable in salads and as a herb to flavour certain Asian foods, especially curry. Some say the taste resembles curried walnuts! Rich in iron and quick and easy to sprout, this is a great way to introduce children to growing herbs at home.

MUSTARD (DRY)

Good source of:
- Sulphur, tryptophan, potassium

Health benefits:
- Contains substances which are antibacterial and fight the growth of fungus.
- Can be used as a stimulant; helps the circulation and fights depression.
- Good emetic; a teaspoon of dry mustard in a cup of hot water will induce vomiting.
- Thought to 'warm' the body and fight depression and lethargy.

Facts and tips:
- Mustard has been used since before the time of Christ and is mentioned in the Bible. One of the most popular and respected condiments, during the fourteenth century the manufacture and sale of its various blends were governed by law.

NUTMEG

Good source of:
- Copper, other minerals and some B vitamins

Health benefits:
- Aids digestion.

Facts and tips:
- Large amounts of grated nutmeg may be slightly intoxicating. It was rated so highly in the last century that gentlemen would carry their own piece of nutmeg and grater with them to their clubs.

  Add nutmeg at the last moment to food, because the aromatic oils containing its distinctive scent are highly volatile.

PAPRIKA

Good source of:
- Iron, potassium, zinc, magnesium and phosphorous, carotene

Also contains:
- Some thiamin, riboflavin, niacin and tryptophan, crystalline ascorbic acid

Health benefits:
- Flavourful way to bring more powerful plant antioxidants to your table.

Facts and tips:
- Paprika is used in Eastern European cooking and often associated with Hungarian cooking. Szeged, in the south of Hungary, is the world's largest producer of this spice.

SAFFRON

Good source of:
- Manganese, also iron, copper and potassium

Health benefits:
Manganese is important for protein synthesis and normal bone formation. Few foods contain significant quantities of this mineral.

Facts and tips:
- Saffron comes from the dried stigmas (male part) of the saffron crocus, which are pulled by hand from the flower and dried: a pound contains about 75,000 stigmas. One of the world's most expensive spices, in the past it was also used as a colouring, a perfume and in various forms of medicine and magic.

MISCELLANEOUS FOODS

BEER (ALSO ALE AND STOUT)

These fermented drinks vary in food content, but tend to be high in minerals, especially potassium and magnesium.

Health benefits:
- Acts as a mild diuretic.
- May help nursing mothers increase their milk flow.
  See also Wine.

CHOCOLATE

(Plain dark chocolate with little else added.)

Good source of:
- Iron and magnesium, a small amount of protein

Health benefits:
- Contains substances (one of which is phenylethylamine) that can improve mood and give a feeling of elation by increasing the amount of serotonin produced in the brain.
- Recent research suggests chocolate contains phytochemicals that help fight heart disease.

Facts and tips:
- Contains caffeine and other stimulants that can affect mood and alertness.
- Substances in chocolate may trigger migraine headaches.
- Chocolate can provide a source of quick energy.

- The best chocolate is that containing 55 per cent or more cocoa. Light creamy varieties may contain as little as 15 per cent cocoa, along with surprisingly high levels of fats and sugars.

GREEN TEA

Good source of:
- Natural antioxidants

Health benefits:
- Green tea contains phytochemicals that help lower blood pressure and fight tumour growth.
- The antioxidants in green tea help fight heart disease and other degenerative conditions thought to be caused by free radical damage.

Facts and tips:
- Natural antibiotics in green tea have been shown to kill the mouth bacteria associated with tooth decay.

MOLASSES (BLACKSTRAP MOLASSES)

Good source of:
- Calcium, potassium, magnesium, manganese and iron; a quick energy source

Health benefits:
- Contain a mix of minerals and vitamins useful to the body's repair and growth.

Facts and tips:
- Molasses are the thick residue from the process of refining beet and cane sugar.

WINE

It holds, pressed from the grape, the secrets of the soil.

Colette

Wine contains no fat, little protein, a limited amount of carbo-hydrate and some alcohol. There is no vitamin C, very few B vitamins and small amounts of potassium. At first there seems to be little to recommend it as a food, and yet, when consumed in modest amounts, medical evidence suggests it has benefit. Some years ago health statisticians discovered a medical phenomenon they called 'the French paradox': here was a population of people living on a very fatty diet, but enjoying a very low risk of heart disease. Experience with other groups suggested the opposite should be true. Was it substances wine contains, called phenols, that protected the drinkers? Or was it the alcohol itself? The debate goes on, but doctors now believe very modest amounts of alcohol – in whatever form – may protect against heart disease.

Other suggested health benefits:
- Dr E. Maury, author of *Your Good Health: The Medical Benefits of Wine Drinking*, suggests that the com-binations of minerals and phytochemicals in specific types of wine have different curative benefits:
- Chablis helps kidney function.
- Dry champagne aids digestion.
- Substances in Beaujolais fight bacterial infection.
- Corbières may relieve some of the discomfort of osteo-arthritis.

Warning:
- Alcohol can be addictive and should be consumed in moderation.

Notes
1 Ferne, A., 'The Great British potato: a study of consumer demand, attitudes and perceptions', *British Food Journal* 94 (1992), no.65, pp. 22–8.

4

QUESTIONS AND ANSWERS
ABOUT HEALTH

The key to health through good nutrition is to eat small servings of many foods. Wholefoods – containing their original balance of nutrients – provide the right combination of natural building blocks needed for growth and healing.

WHICH FOODS FIGHT CANCER?

Plan meals to include at least two of the following foods in your diet each day:

Star Foods

soya beans and soya protein
Shiitake mushrooms
garlic
all citrus fruits
apricots
tea (some research shows green tea to be of greatest value)
cruciferous vegetables – broccoli, cabbage, Brussels sprouts,
    cauliflower
carrots, red peppers
parsley
Spirulina

Basic Foods

berries, particularly those with a red colour (cranberries, red
    currants, blackberries, strawberries)

pink and orange fruits, including melons, persimmons (Sharon fruit), mangoes and papaya

red and green vegetables, including pumpkins, spinach, yams and turnips (young turnip tops (greens) make an excellent side dish)

radishes

grains – whole wheat and wheatgerm, barley, oats

nuts – walnuts, almonds, and hazelnuts

seeds – pumpkin, sesame and sunflower

pulses – chickpeas, lentils, dried beans

herbs and spices – camomile, cloves, sage, tarragon

monounsaturated and polyunsaturated oils – extra virgin olive oil, walnut oil (use only good quality oils and make sure they are not rancid)

foods high in natural fibre such as pulses, wholegrain foods, fruit and vegetables

Are there foods I should avoid?

Reduce your total fat intake to 30 per cent or less of the total calories you consume, and avoid oils and nuts that are rancid. Also avoid mouldy nuts and grains, as they may be contaminated with aflatoxin, a powerful natural carcinogen. Alcohol abuse adds to the risk of cancer. Adjust your diet to avoid highly salted and barbecued foods.

Are organic foods important?

Organically grown foods are less likely to contain harmful chemicals and should be our foods of choice whenever possible. It is not always possible to avoid chemical contamination, however, due to previous soil usage and environmental pollution, so make certain you wash all foods well and, unless you are absolutely certain of their source, peel root vegetables.

What about meat?

This is not a book about eating meat, but it needs a mention here. If you can, buy eggs, milk and milk products, and any meat products you use from reputable organic sources.

Do any foods promote cancer?

To 'promote' cancer, a substance needs to help the action of a carcinogen, or cancer-inducing substance. Evidence suggests that alcohol enhances the cancer-producing effects of aflatoxin and enhances the effect of substances thought to be associated with cancers of the oesophagus.

High-fat and low-fibre diets are associated with an increased risk of breast and colon cancer, and a high-fat diet is a factor in the development of pancreatic and prostate cancer. Low intake of fruits and vegetables and high intake of highly salted food appears to increase the risk of stomach cancer.

For several decades, scientists have suspected the charred material on barbecued foods may contain carcinogens.

Does it matter how food is preserved?

Yes. The high intake of foods preserved by salting and with substances containing nitrates is linked with increased rates of cancer. Freezing and canning require few, if any, preservatives and are safe ways to preserve food. (Some powerful synthetic antioxidants used to preserve foods have been shown to reduce the risk of cancer in animal experiments.)

WHICH FOODS FIGHT HEART DISEASE?

Heart attacks and cardiovascular degeneration due to atherosclerosis are diseases of Westernized societies. The degenerative processes which cause these illnesses can be slowed or prevented by changing the way we eat. Your daily diet should include at least three of the following star foods and two of the basic foods:

Star Foods

garlic and onions
cabbage, broccoli, kale, turnips and cauliflower (*Cruciferae*)
oats

extra virgin oils, olive and walnut in particular
tomatoes
citrus fruits (lemons and limes are particularly good)
all red and green vegetables and fruits (pumpkin and apricots,
    for example)
basil
dried beans
nuts and seeds
oily fish – mackerel, salmon and fresh tuna are good
artichokes (globe)
Spirulina

Basic Foods

grains – barley, buckwheat, rice, maize (corn), whole kernels
    and used as polenta
herbs – thyme, marjoram, chervil, chilli and watercress
fruit – strawberries, cherries, grapes, plums, melons
root vegetables – potatoes, yams
other vegetables – chard, pumpkins

Which foods are good for my heart?

There are two main things to remember when thinking about
food and heart disease. First, we need to choose foods low in
saturated fats but rich in essential fatty acids. Second, excessive
quantities of free radicals caused by many environmental factors can be deactivated by a steady supply of natural antioxidants from the foods we eat.

Around the mid-1980s food and health writers began talking
about the Mediterranean diet as though it were something
newly discovered, although the low rates of heart disease
among people from this region had been scientifically recognized for about 30 years. Through being bombarded with arti-
cles about garlic and tomatoes, oily fish and extra virgin olive
oil, however, many of us rediscovered the joy of eating a good
pasta, or the delights of a fine salad made of ripe tomatoes, fresh
basil and black olives served with hot bread rubbed with garlic.
Increasing the amounts of these wonderful foods in our diets

not only supplies our bodies with the natural nutrients needed to fight illness and maintain good health, but also demonstrates how easy it is to eat well-balanced and delicious meals based on foods derived from plants.

The food and health writers have not told us the full story, however: good Mediterranean cooking is not only rich in fruit, vegetables, herbs and fine olive oils, but also contains cabbage and green vegetables from the cruciferous group of plants, nuts (including pine nuts), oils from nuts (walnut oil is especially good), parsley and dried beans.

What is the most important rule when fighting heart disease?

Be aware of your heritage. If you have a family history of early death from coronary illness, pay particular attention to your diet and lifestyle. Do not smoke, cut down on saturated fats, exercise in moderation, base your diet on the delicious foods listed above and enjoy life: stress and weakened hearts are not healthy companions.

What about salt?

Watch the salt! Salt is such an important part of our diet people have fought over it and used it to pay soldiers their wages. However, excess salt in food raises blood pressure and that increases danger from strokes and heart attacks. Skip the salted nuts and crisps (potato chips), and let the salt shaker on the table take a rest for a while. Our taste buds adjust to the amount of salt we use; stop salting the food on your plate and you will soon enjoy delicate and unique flavours you may have been missing!

We do need salt in our food to balance our internal chemistry, however, so unless your doctor advises you to cut out all of the added salt in your diet, there is no reason to avoid salt when cooking. You can't make many basic dishes – like breads, soups and stews – without it. When possible use sea salt; it has a fuller flavour than plain sodium chloride crystals because it also contains a wealth of other minerals our bodies require for good health, such as magnesium and iodine. If you live in an area where the natural levels of iodine in the soil are low and you eat little or no seafood, make sure you choose iodised salt. Iodine is essential for normal growth during childhood and for a long and healthy life.

What about sugar?

Some people in the Western world consume massive amounts of refined sugars in soft drinks and confectionery; this throws their metabolic systems out of balance and opens the door to obesity and related problems. Be sensible about sugar – enjoy natural sweetness from fruit for most of your desserts and energy snacks.

What about fats?

Enjoy foods from plants containing unsaturated fats. Avocados, green leafy vegetables, seeds and nuts are good. Watch out for coconut oil, however, as it is high in saturates. Also, remember that oils are fluids because they contain molecules with the delicate unsaturated carbon bonds we need for good health. When a food manufacturer changes liquid oil into a spreadable solid, something has been done to change the characteristics of the oil and probably the beneficial parts of its basic structure.

WHICH FOODS STRENGTHEN THE IMMUNE SYSTEM?

The immune system not only helps fight infection, it also helps fight the development and spread of cancer cells. To strengthen it, your diet should include at least three of the following each day:

Star Foods

garlic
foods rich in vitamin C – strawberries, citrus fruits, ripe
 tomatoes

oils and nuts rich in vitamin E
foods rich in beta-carotene and other carotenes – red and dark
    green vegetables
foods rich in zinc – pumpkin seeds
herbs – ginger and thyme
Spirulina

Basic Foods

fortified cereals
legumes – beans, lentils and peas
fresh fruit and leafy vegetables

Which mineral is most important in helping my immune system?

Zinc is a mineral micro-nutrient specifically needed by the cells
of the immune system. Lean red meat is a good source, but if
you choose not to include this in your diet, make certain you
take mineral supplements to improve your immune system.
Elderly people particularly benefit from supplements. *(See page
80 for plant sources.)*

What about vitamins?

Vitamin B_{12} is very important to a healthy immune system; studies have shown that low levels of this vitamin increase the like-
lihood of contracting infections, including tuberculosis, a
serious infection that is on the rise in Western countries where
it was once thought to have been eradicated.

To make sure your diet contains adequate vitamin B_{12},
include yeast extracts or Spirulina in your diet each day.

Are there foods I should avoid?

Yes. Reduce the quantities of saturated fat and processed sugar
in your diet, cut back on the number of cups of coffee you drink
each day and limit your alcohol intake.

Are there times when I need to be particularly aware of the effect diet has on my immune system?

If you have had a long illness or suffer from an immune-related illness such as arthritis or multiple sclerosis, modify your diet to include the foods listed above.

WHICH FOODS HELP SLOW AGEING?

Nothing will *stop* ageing – it is the result of a series of natural processes built into the basic structure of our genetic control mechanisms – but the effects of ageing, including skin wrinkles, cataracts, rheumatoid arthritis and atherosclerosis, can be *slowed* by enjoying foods rich in powerful antioxidants. Most important among these are the natural vitamins C and E, and beta-carotene (the substance from which the body makes vitamin A). These, in combination with the mineral selenium, interact with and deactivate rogue molecules, called free radicals, which can rip across and destroy the biological activity of delicate molecules needed for normal cell and tissue structure.

Star Foods

all brightly coloured red, orange and green vegetables and fruits

citrus fruits, strawberries, potatoes

wholegrain cereals and nuts, particularly walnuts, which are also a good source of essential fatty acids

wheat bran

Spirulina

soya protein

Some tips on fighting ageing:

* Think young! Eat young! Build your diet around the bright colours of fresh fruits and vegetables, nuts and seeds. For protein and phytochemicals useful in preventing prostate and breast cancer, include about 50 grams (1.8 oz) of soya protein in your diet each day.

* Choose foods that are low in saturated fats. Use food incorporating natural oils instead. Enjoy foods that are naturally low in fat to help keep your total fat intake below 30 per cent of your total caloric intake.

CAN EATING FOOD FROM PLANTS HELP REDUCE BODY WEIGHT?

Yes. On their own they are low in fat and high in nutrients needed for healthy tissues.

Fruit, vegetables and foods made from grains and nuts are also good sources of natural fibre, which helps you feel full. And they are rich in complex carbohydrates, which break down slowly in your digestive system and control the sensations of hunger. If you suddenly do feel hungry, or if your appetite has been so roused you *must* find something to eat, enjoying a piece of fruit or a few sticks of crisp, fresh vegetables will help get things back on track. You can enjoy the satisfaction of chewing and reap nutritional value from your snack at the same time.

But remember, by cutting down your total food intake you reduce the amounts of specific nutrients available for your body to carry out the various processes of life. Recent research at Oxford University suggests that restricting food may result in low levels of tryptophan, an amino acid which the brain needs to produce an important chemical messenger, serotonin (also called 5-HT). Among its various tasks in the brain, serotonin lets it know when we have had enough to eat by linking with chemical structures called 5-FHT receptors. It is thought that when tryptophan levels are low and the 5-FHT receptors are not satisfied, hunger messages continue to bombard the body. Once activated, the hunger drive can result in discomfort and over-eating.

What can you do about this? Include tryptophan-rich foods in your diet. Bananas are an excellent source.

WHICH FOODS HELP CONTROL...

ANAEMIA

Enjoy foods rich in iron, including green plants and fortified cereals. Reduce your tea intake, because substances in tea block iron absorption.

ASTHMA

This crippling lung disease is on the increase among both children and adults. A mark of our time was recently seeing a youngster in a fast-food restaurant sitting in a special chair and fitted with breathing equipment supplied by the chain as a service to its customers! What causes this debilitating plague? Some experts blame an increase in air pollution; others claim it has something to do with diet.

A link has been identified between eating fresh fruit and vegetables and asthma; as children fail to eat adequate quantities of these foods, the risk of asthma increases.

CANDIDA ALBICANS

Unlike many other, helpful, micro-organisms that populate the human body, the yeast *Candida albicans* is an unwelcome inhabitant of the digestive and reproductive tracts because it can cause oral and vaginal thrush, cystitis, headaches, fatigue, diarrhoea, constipation, anxiety and depression.

Load your diet with green and yellow fruits and vegetables; they are full of nutrients needed to build up your immune system and strengthen tissues and are a rich source of phytochemicals that fight infection. Shift your carbohydrate intake from simple to complex carbohydrate, skip the sugar on your breakfast cereal, enjoy an evening meal of bean soup or a salad made with wholemeal pasta. And finally, for a healthy intestinal system, enjoy oats, walnut and green leafy vegetables rich in omega-3 fatty acids. Research suggests too much saturated fat and fried foods lower the body's resistance to yeast infections. The omega-3 fats support the fight against infection.

For more information, read *Candida Albicans: How your Diet Can Help*, by Stephen Terrass (Thorsons, 1996).

CEREBRAL ACCIDENT
See Stroke

DIGESTIVE PROBLEMS: LIVER, PANCREAS AND 'DELICATE' STOMACH

To aid digestion, enjoy more apples, bananas, pears and dark coloured berries (raspberries, blueberries, etc.), artichokes, asparagus, wholegrain pasta, and foods made with oats, brown rice and walnuts.

If you are suffering liver problems that result in jaundice, or yellowing of the skin, eat plenty of soya protein to help provide healthy materials for tissue repair, and eat foods rich in iron and B vitamins, especially Folate. Wholegrain foods and yeast extracts provide needed nutrients. Ask any Frenchman and he will tell you artichokes are one of the best foods for an over-worked liver or the effects of too much wine.

If you suffer from frequent indigestion or ulcers, concentrate on eating a diet rich in green and yellow fruits and vegetables. These are rich in vitamin C and beta-carotene. Also eat plenty of foods rich in zinc, such as wholegrains. Cut back or eliminate caffeine from your diet and look with suspicion on salty and spicy foods, which can irritate ulcers.

DRY SKIN
See Eczema

ECZEMA AND DRY SKIN

Reduce the saturated fat in your diet and enjoy more foods containing omega-3 fatty acids, such as oats and walnuts.

FATIGUE

Fight fatigue by supporting your body's production of red blood cells. Red blood cells carry oxygen to all the body's cells. Enjoy nuts and green vegetables to up your intake of Folate. Make sure you eat foods high in iron, vitamin C, zinc (to help the enzymes in energy metabolism) and vitamin B_{12}. If you are a vegetarian, you may feel tired because your B_{12} levels are low; a dietary supplement may help.

Cut out the sugary snacks – they do terrible things to your blood sugar, making it peak and fall in rapid succession. And they do nothing for your teeth! As your blood sugar level swings up and down, so does your mood. Have a handful of sugary sweets or candy and you have a sugar 'high', but in about half an hour, when the blood sugar falls, you feel tired and start looking for another quick pick-me-up. Energy from complex carbohydrate, on the other hand, is released slowly into the blood, and keeps you on an even keel for longer. Next time you feel tired and crave food, try a snack high in complex carbohydrate. An oatmeal biscuit, a handful of raisins or a few pieces of dried apricot or a slice of crisp bread may be just what you need.

If you feel tired, you may need more iron in your diet. Good plant sources include green leafy vegetables, lentils, nuts, wheatgerm and sunflower seeds.

Stress and poor sleeping habits often go hand in hand. *See the section on foods that help fight stress (p.190).*

GALLSTONES

There is a school of thought that extra virgin olive oil helps relax the gall-bladder and encourages the release of bile into the gut. Some people achieve this therapeutically by occasionally drinking a considerable quantity of olive oil, interspersed with lemon juice. The effects are rather unpleasant and dramatic. However, people living in some parts of France drink olive oil as if it were a fine whisky. As many people have gallstones without ever knowing they are there, an occasional emptying of the gall-bladder may help avoid the build up of a potential problem. The best approach would be the frequent use of a vinaigrette made with virgin olive oil and lemon juice.

HIGH BLOOD PRESSURE

If you need to reduce your weight, cut calories by avoiding foods rich in saturated fats. Add soya beans or soya products to your meal plans (tofu, miso and soya milk are versatile foods). Watch the salt; cultivate a taste for raw vegetables without dips and dressings. Eat to keep your blood sugar level steady to avoid mood swings and hunger pangs: skip the fat and enjoy starchy foods such as bread, pasta, cereals, potatoes, lentils, beans and nuts. These are rich sources of fibre and the essential nutrients needed for good health.

HIGH CHOLESTEROL LEVELS

Besides enjoying more Mediterranean foods, plentiful in olive oil, garlic and antioxidants, to avoid high cholesterol levels, eat more foods rich in essential fatty acids and add soya products to your diet.

MOUTH ULCERS

Aphthous ulcers, painful white indentations in the soft tissue of the mouth, can result from a diet deficient in zinc and the B-vitamins, although stress, hormonal changes and food sensitivities also have an effect. If you suffer from recurring mouth ulcers, consider increasing the quantity of dark green vegetables in your diet and enjoy more nuts and other foods with a high zinc content.

Mouth ulcers are more painful when exposed to salty or acidic food and to alcohol.

Warning: Any white patches in the mouth that do not heal in a week or 10 days should be examined by a doctor, as they may be an early sign of cancer or other conditions requiring medical attention.

OSTEOPOROSIS

Although osteoporosis (thinning of the bones) is considered a disabling condition affecting many post-menopausal women, it is also a condition suffered by men. Historians claim Oscar Wilde died of complications from osteoporosis, for example. Dancers, certain types of athletes and people on starvation diets may also be sufferers of this painful condition.

A good way to prevent this disease is by consuming foods rich in calcium and vitamin D throughout your lifetime *(see pages 52 and 69)*. Many doctors prescribe HRT (Hormone Replacement Therapy) to help reduce the risk of osteoporosis; recent research suggests the phytoestrogens in soya beans and certain yams may be an effective alternative to HRT. Reducing consumption of salt, alcohol and caffeine may also protect bone density.

'RESTLESS LEGS'

This is a common condition which receives too little attention. Sufferers experience unpleasant tingling, muscle jerks and burning sensations on the skin of the thighs and calves.

Diet changes that may help include reducing your intake of caffeine, especially from coffee, and eating more foods containing potassium (bananas are a good choice), iron, Folate and vitamin E.

SPINA BIFIDA

A link has been proved between low levels of Folate (Folic acid) in the mother's diet and spina bifida, a crippling deformation of the spinal cord in their infants. Preferably before you conceive, but certainly when you are aware you are pregnant, increase the amount of Folate-rich foods in your meals. Enjoy more broccoli, Brussels sprouts, pulses, barley, soya, rice and wheatgerm, and treat yourself to foods containing seeds and nuts. If you enjoy yeast extract, add that to your shopping list. When you know you are pregnant, see your doctor and ask whether you need a Folic acid supplement.

Make sure your average food intake is sensibly balanced to provide all the nutrients needed for general good health. Focus your food choices on those rich in minerals and vitamins B complex. Skip products made with refined (white) flour and add foods rich in zinc, calcium and magnesium, including wholemeal bread, nuts (especially pecans and walnuts), blackcurrants, broccoli and melon. Also add foods with a high tryptophan content. This essential amino acid is found in soya beans, sesame seeds, walnuts, bananas and pasta. Tryptophan increases the brain's production of serotonin, a chemical that gives a feeling of well-being and induces relaxation. Low tryptophan diets have been associated with aggressive behaviour, which can be reversed by dietary improvements.

When combating stress, avoid colas, coffee, tea and other foods containing caffeine. Also avoid the temptation of eating a chocolate or candy bar for a quick sugar 'high' – as any sugar-junkie will tell you, that high will be followed by a 'low'.

In his book *Stress: Proven Stress-Coping Strategies for Better Health* (Thorsons, 1995), Leon Chaitow describes how to develop your own 'stress protection plan' and why tryptophan can help you stay calm under pressure.

STROKE (CEREBRAL ACCIDENT)

Concentrate on enjoying foods high in antioxidants – nuts rich in vitamin E and strawberries and potatoes rich in vitamin C, for example. Diets with a high fruit content can reduce the risk of stroke, which is the third most common cause of death in the United States. Research at the Harvard Medical School, published in the *Journal of the American Medical Association* in 1995, demonstrated a 22 per cent drop in the risk from stroke when three or more servings of vegetables and fruit were included in the daily diet. Vegetables were found to be more effective than fruit.

THYROID FUNCTION

The thyroid is a small gland in the front of the throat that secretes a hormone important in regulating the body's rate of metabolism. If the rate is too high (hyperthyroidism), try increasing the wholegrains, seeds and pulses in your diet. If it is too low, add seaweed and use iodised salt in cooking.

CAN EDIBLE PLANTS CAUSE PROBLEMS?

Yes, there are a few:

* Watch out for nut allergies. If you are a sufferer, avoid both nuts and unrefined oils.
* The nightshade plants are vegetable favourites that can be less than friendly to some people. Although they appear to have little in common when viewed in your shopping basket or on the plate, red and green box peppers, aubergine (eggplant) and tomatoes are all from the same plant family: nightshade. The fruits of this group contain high levels of alkaloids, which may be implicated in the development of rheumatic symptoms and increased heart rate. Because alkaloids are also known to reduce the rate of absorption of B vitamins, they may all add to stress levels. B vitamins are known to help control emotional stress, so if you enjoy lots of tomato sauces and potato salad, you may want to think about making sure you also lace your diet with vitamin B-rich grains and seeds.
* Calcium and iron-rich foods are vital in a healthy diet. Reduced calcium levels can increase the risk of kidney stones and osteoporosis. Unfortunately, some plants contain high levels of oxalic acid, which binds and removes calcium from both food and body tissues. Foods containing high oxalic acid levels include spinach, rhubarb, beetroot tops, Swiss Chard and buckwheat. Do not eat these plants in the same meal as foods chosen for their high calcium content.

5

TIPS ON BUYING AND USING
FOODS FROM PLANTS

1: HEALTHY COOKING WITH
FRUIT AND VEGETABLES

All good cooking begins with the basic rules of kitchen hygiene. Make sure everyone in your family knows that the first rule is: *Wash your hands*. Second: Don't smoke in the kitchen; cigarettes and food do not mix. (In fact, for the sake of your health, don't smoke at all!) Third: Keep pets out of the kitchen. This is not the best place for Rover's bed. And fourth: Remove rings and watches while you cook. You would be amazed what creeps in under your rings or hides in the little spaces between the stones.

Further tips:

* Keep a separate cutting board for use only when you are preparing fruit and vegetables. The surfaces on which these foods are sliced should never be shared with fish, poultry and red meat. Bread should have its own board and knife.
* Store fruit and vegetables in a dry, well ventilated place when you get them home. In the kitchen, place them in containers designed to allow the maximum circulation of air around the food. Baskets are good, as are some of the vegetable trolleys and hanging wire containers on the market. Woven or 'punched' ceramic bowls are ideal for delicate peaches, grapes and soft fruit.

    Place foods that require refrigeration in the covered boxes at the bottom of your refrigerator. The temperature here is slightly higher than elsewhere in the refrigerator and better for fresh produce. Although it may not be obvious, you can be sure the outside of the food is covered in various kinds of chemicals and dirt: that is an expected part of the picking, packing and handling processes needed to get food from the field to your home. By their very nature, vegetables – particularly the leafy varieties – carry dirt and bacteria on their surfaces. But do not wash these foods at this time, as introducing moisture to their surface may increase the growth of normal bacteria and increase the speed of decomposition.

(The healthy way to store foods in a refrigerator is to arrange covered, cooked foods at the top, where nothing can drip on them, and uncooked foods below. Fish or meats should be placed in containers and not allowed to come into contact with cooked foods. Notice that there is usually a glass shelf separating the bottom vegetable boxes to protect them from dripping foods above.)

* Just before using, wash all fruit and vegetables under running water and scrub with a vegetable brush to remove pesticide and fertilizer residues. Washing also cleans off any dirt from soiled hands. It is easy to get excited over chemical traces and forget the grime and bacteria deposited by busy hands as they harvest and pack our food. If you are preparing fruit with edible skins, such as apples, you may want to add a few drops of washing-up liquid (dish soap) and scrub the surface with a vegetable brush under lots of running water. Unfortunately, tests have shown the skins of some root vegetables – particularly carrots – can be contaminated with farm chemicals; peel to be on the safe side. Teach your children to always wash fresh fruit well before enjoying that first juicy bite.

* Wash leafy vegetables – lettuce for example – in running water to remove sand and dirt. Remove and discard the tough outer leaves. Dry on clean towelling before use, wrapping and gently pressing to remove any excess water. Again, do this just before use, because the processes of washing and drying will bruise the leaves and encourage bacterial growth.

* Use filtered water to cook vegetables, grains and pulses, and to brew tea. (A simple jug fitted with a charcoal filter is fine.) This eliminates some of the impurities in the tap water that can affect the flavour of food and drink. Always use filtered

water to soak beans and other pulses (legumes) overnight, as this water is absorbed into the food and becomes part of what you will eat later. Just remember that jug filters trap bacteria that may reproduce in the charcoal; change the filter at the intervals prescribed on the package and keep the jug in the refrigerator, top shelf if possible.

* Cooking extra portions of rice and pulses to use in a later meal makes good sense, but make sure you refrigerate the excess as soon as it reaches room temperature. Leaving high carbohydrate foods overnight or longer at room temperature encourages the growth of mould spores that may lead to stomach upsets later.

* All beans should be soaked overnight before cooking, but red kidney beans need extra attention before use. If not fully removed, certain natural chemicals in the beans can lead to stomach discomfort. Rinse and soak the beans, cover with twice the level of water needed to cover the beans, soak overnight (at least eight hours) at room temperature, pour off the excess water and rinse, cover with water again and cook for two hours, topping up with additional water when necessary. Do not salt or add stock cubes to the water during this early stage of cooking, because this toughens the outer shell of the beans, making them difficult to digest.

* At parties and other social gatherings, do not put out large quantities of food and leave them for hours. Put out smallish quantities and change them regularly. This does not mean topping up a small dish – it means removing a bowl or dish and replacing it with a fresh, full one.

HOW TO MAKE A PERFECT SALAD

* Buy a salad spinner; it is worth the money. All crisp salad greens should be washed before using, and trying to remove the excess water by shaking or rolling in a kitchen towel bruises the leaves and leaves them limp. A twirl in a salad spinner leaves the greens in perfect condition to hold a light coating of dressing.

* Buy fresh salad greens. This is one place where freshness really counts. Choices will vary with the season, but buy

enough of what looks fresh and crisp to cover the bottom of your salad plate; remember, salad gives colour and height to a dish, so you want colour and height in the salad greens you buy. A little chopped celery also adds crispness to a salad.

* Add some weight to your salad. Cooked and chilled chickpeas or diced root vegetables are ideal. Go wild! In addition to the carrots, potatoes and beetroot, try turnips, parsnips, sautéed mushrooms, water chestnuts or lightly fried chunks of red pepper, aubergine or courgette.

You may want to try building a salad on a bed of cooked, chilled and seasoned grain: a bed of tubule or kasha or a slice of fried polenta can form the base for a salad of mixed vegetables.

* Dress the salad greens and firmer vegetables separately. Just before serving, make certain the greens are dry, then coat with a light dressing of oil and vinegar or lemon juice and season with salt and pepper. Arrange these in the centre of the serving plates. Dress the heavier ingredients with a heavier dressing, a mayonnaise or a mixture of extra virgin olive oil and lemon juice for example, and arrange in the centre of the plate on top of the salad greens. (A small spoon of wholegrain mustard or a spoonful of ketchup added to an oil dressing can raise the flavour of the salad.)

* Add texture. Most people like crunchy food; take advantage of this and top the salad with a good pinch of toasted nuts or seeds to add valuable nutrients as well as eating quality.

* Dressing brings it all together. Homemade is fresher, better and cheaper. The easiest way to make a vinaigrette dressing is to place all of the ingredients (use about three times as much oil as vinegar or lemon juice) in a screw-top jar and shake.

2: CUDDLE A RADISH? HELPING CHILDREN ENJOY FOOD FROM PLANTS

Giving our children a positive attitude towards healthy eating can be one of the most important things we do to protect their good health in later years. Unfortunately, a common carrot or bunch of grapes has a hard time competing with the appeal of a well designed wrapper on a chocolate bar. So we need to show children that fresh foods are interesting and – most important – delicious. This is no easy task, as it requires us to give children opportunities to make choices on their own, explore and try things that might be slightly over the top! Perhaps the following suggestions will inspire you with ways to encourage your children to enjoy more fresh foods.

* Use pictures of fruit and vegetables as part of the decoration around your home and in the areas where your child sleeps and plays. In addition to pictures of friendly rabbits and happy ducklings, how about a few smiling chunky cherries or some lovable radishes?
* Let children see you enjoying foods made with fruit, vegetables, grains and pulses. It is no good for a dad to encourage his son to eat up his broccoli, only later to push it to the side of his own plate! Bear in mind, though, that this may not always work – as my copy-editor said, 'Although my mother has always eaten spinach and marrow with a smile, I know I will never manage to force them down!'
* When your child is old enough to eat soft foods, introduce a range of flavours; parsnips instead of carrots, for example.
* Try making you own puréed vegetables at home; in addition to introducing your child to a wider range of tastes and textures, your own blends will have more flavour, fewer additives, more nutrients and cost less.
* As your children get older, try to understand what they think about various foods. Holding an impromptu art contest on a rainy day can provide interesting insights for both your children and yourself. Supply paint and paper and ask who can draw the best cartoon of a meeting between a food they like and the food they think should be banned from the dinner table! The results may surprise you. See what you can learn and try adjusting the way you prepare certain foods. Let your family know you are trying some new food ideas, based on the cartoon contest, and get back some responses.
* Teach your children to cook. Begin with salads and move on to soups. Marion Cunningham, an American who has taught

children to cook for more than 20 years, believes children under the age of 10 should be able to prepare the family dinner.

* Have a *Food Search*. For children old enough to be let loose in the supermarket (find a quiet time of the week to do this!), make a short list of two or three healthy foods you know they enjoy and send them off to find them. Give instructions to bring back one container of each item on the list, plus one other item they would like to try from the same shelf. So a can of peaches may be partnered with a can of pitted cherries or a can of figs, for example.

* Become *Food Detectives*. While you do the regular food shopping, let your children (even as early as three or four years) select one item a week to try. Steer them away from the sweets (candies) and savoury snacks, however, except for very special occasions like Christmas or a birthday. Then, once the groceries are unpacked, let your child handle their selection. All too often there is a long time lag between what comes out of the store and what goes on the table. Make the association between selecting and exploring the food more immediate. Explore with your child. Depending on the food selected, smell it; cut it open; try tasting some raw; cook some in boiling water; fry another slice; sprinkle it with lemon juice, and so on.

* Have a *Food Month*. Make every Saturday lunchtime for one month special by letting children dream up their own combinations of textures and flavours. Supply them with healthy ingredients – bread, hard boiled eggs, peanut butter, canned sardines, mustard, a bowl of sweetcorn, a sliced tomato, broiled bacon, raisins, apple slices, and so on. Provide large plates and lots of space while the kids take over the kitchen, then get out of their way. There are only two rules: they eat what they prepare, and they clean up after lunch. At the end of the month, let them make lunch for you and talk about what they have discovered.

* Encourage healthy choices. Children need fat in their diet; the trick is to help them enjoy the fats that are best for them. For example, at dinner encourage children to try a variety of oil-based dressings on salad and cold steamed vegetables

served with a slice of chicken or a piece of cheese. Make, or buy, some savory vegetable sauces and pastes – tapenade, pistou, hummus and guacamole for example – and warm a loaf of bread in the oven and put the whole lot on the table for an afternoon snack.

* Watch their garden grow! Even if your gardening is limited to a few clay pots on the back terrace there is room for growing one or two herbs. Parsley – that fantastic source of antioxidants and other natural healers – grows with very little attention other than some water and a few hours of sun each day. Let your children tend the herbs and 'harvest' them for the foods you prepare. Then give them a chance to try their hand at mixing a few herbs over the boiled potatoes or on top of a piece of fish before it is served.

3: HEALTHY EATING ON A BUDGET

Food is a major expenditure in most households and, like any other investment, when you spend money you want good value. You are looking for well-priced food with two benefits: it should appeal to your hungry family and contain all the nourishment they need for good health. To help make your money go further and to bolster the goodness in the food on your table, try these tips at home and when you are shopping.

AT HOME

* Buy foods in season. It is great to shop for strawberries in the dead of winter, but the cost hits the pocket where it really hurts. In cold weather, enjoy root vegetables. They are full of vitamins and minerals that support the immune system.

* Look for cheaper substitutes for healthy foods. For example, I despair over both the quality and price of tomatoes in the supermarkets during the cold months of winter and spring. Why must we pay good money for such poor texture and taste? So at this time I depend on the deep rich flavour of tomato paste, sun-dried tomatoes, or succulent tomatoes from a can! Even in salads, sliced, firm plum tomatoes add

real flavour and beautiful colour. Just drain them well and be gentle when you slice them. The lovely juice which accompanies them can be used in a sauce or soup – or drunk as a cook's quick pick-me-up.

* Go ethnic! Many regional foods, which have been the source of good health and long life for many people over the centuries, are simple, highly nutritious, inexpensive and based on plant and plant-derived foods. Just look at the Mediterranean diet – where would Italy be without its pizza and pasta?

* Buy locally grown foods when possible. Is there a cooperative in your area? There are many in the United States.

* 'Crop share' – if you and a friend enjoy gardening, make an agreement to trade.

* When planting your spring hanging-baskets, mix in herbs with the flowering plants to enjoy their scent and flavour all summer long. Good choices are parsley, marjoram and rosemary. Fresh mint is a real boon to summer cooking and it is easy to grow. As already mentioned, though, it is best to plant this herb in a large pot and then place it in the ground. Mint can become a weed if its roots are not kept under control.

* There is a place for frozen food in a healthy diet. It is often cheaper and easier to buy frozen vegetables than fresh. These can be a boon to a busy cook. Frozen foods may lack some of the crunchy texture available in fresh produce, but most of the vitamins are intact. Some frozen vegetables contain more vitamins weight for weight than their fresh counterparts.

Super Tips

* Cabbage is particularly rich in associate nutrients, but is sometimes cumbersome and time-consuming to prepare. In frozen form, however, it is highly versatile. Steam or lightly cook some in a pan with a little oil, add salt, pepper and a squeeze of fresh lemon juice, top with a few drops of good olive or nut oil, sprinkle with a few plumped raisins or toasted sesame seeds, and serve.

* Alternatively, try using frozen cabbage in soups: take equal portions of frozen cabbage, thinly sliced onion and cold

potatoes left over from yesterday's dinner, sauté the onions and cabbage in a little oil and when they are translucent add the potatoes, mashed or cut into pieces, and fry gently until lightly crusted. Turn and fry on the other side. If you want some colour and the addition of another vegetable to your meal, add half a finely chopped red or green pepper along with the onions and cabbage. Topped with a poached egg, this makes a healthy and inexpensive meal.

* Make your own foods for your freezer. Instead of using expensive commercial frozen 'quick meal' dishes, make a double or triple batch of your family's favourites, cool extra portions, wrap well, label and store in the freezer for up to one month. Freezing is an excellent way to preserve herbs: pick, clean and finely chop leaves, partially fill ice-cube trays with filtered water and spoon chopped leaves into each division within the tray. Top with more water to fill. Freeze for at least 12 hours, empty ice cubes into labelled freezer bags and store. This will keep you cooking with fresh herbs for up to three months. Use by simply dropping cubes into the food you are preparing, or (less effective) melt herbs in a fine sieve and then add to food.

* Remember, saving money depends on clever budgeting of assets and what you have in your freezer is an asset. But the longer food is stored, the more likely it is to lose its texture and true flavour. Always keep a list of what you have on hand and work your meal planning around that, using the oldest or most delicate foods first. Discarding food because it has been stored too long is a terrible waste.

* Make your own *bouquet garni*. Use cheesecloth (muslin) cut into 4 or 5 inch (10 or 12 cm) squares. In the centre of each, place a bay leaf, two or three sprigs of dried thyme, some fresh parsley and a some rosemary or sage. Tie the cloth over the herbs with thread or fine twine, place in a freezer bag and label. Use within two months. When you take a *bouquet garni* from the freezer, rub it between your fingers to break the leaves and release the flavour.

* Things do go wrong. If vegetables are overcooked, don't fret. Place them in a blender, whiz for a few seconds, add some oil or butter, season to taste, top with some toasted nuts or seeds

THE PLANTS WE NEED TO EAT

and serve a beautiful purée to delight your family or guests!

* Leftover vegetables can form a nutritious part of a stew or soup, or a stuffing in an omelette. Plan ahead: vegetables served hot one night can be chilled and used as part of a vegetable salad the next. For best texture and taste, use leftover vegetables within 24 hours.
* Enjoy making vegetable soups and stews.

How to make a quick fresh pea soup:
(This same method can be used for any vegetable, but frozen peas make a most wonderful dish.)

Ingredients:
500 grams ($1/2$ lb) ($1^1/2$ cups) frozen peas
(petit pois give the sweetest flavour)
450 ml ($3/4$ pint) vegetable stock or cream
fresh herbs
salt and freshly ground pepper

1 Simmer the peas in the minimum amount of vegetable stock until they are cooked, but still have a firm feel when pressed between your fingers.
2 Pour stock and peas into a blender and briefly whizz once or twice until a thick purée has formed. For added interest, blend in some freshly washed mint leaves, or some basil. Thin with more vegetable stock or some double cream. Taste and season with freshly ground pepper and salt.

This soup can be served hot or cold.

WHEN YOU SHOP

* Before shopping, check out the refrigerator and freezer to see what is on hand. How many times have you bought a bag or box of this or that, only to find you already have two at home?! Also, make a list of what you think would be good main dishes for the week ahead and use this as a guide when you buy. Don't be too rigid about sticking to the list, however, as a sale or selection of well-priced items may inspire you to change your plans.

✱ As you shop, remember that the store management is there to separate you from as much of your money as possible. Market research tells them that a good place to put expensive items is at eye level, where you look first. Less expensive items, usually with the same nutritional value, are often placed on top or bottom shelves. So to save money on food, be prepared to look up and down and reach for what you really want.

✱ Organize your shopping trolley (shopping cart) to help you see the balance between healthy foods derived from plants and processed foods and foods high in fats. At the far end, pile up the fresh fruit and vegetables. This should be a colourful display of green, yellow and red by the time you have worked your way through the fresh food section. Remember, yellow and orange fruit and vegetables, including yams, carrots and peppers, are all good sources of the vitamin A, beta-carotene and fibre needed for good health.

Vegetables and fruits sold loose are often cheaper than those pre-packaged and you get to choose exactly what you are going to take home.

Next, head for the frozen food section. Place bags of frozen vegetables – spinach, peas or cauliflower for example – next to the fresh items. Frozen vegetables are usually excellent value for money and mixed vegetable blends make it possible to serve your family a variety of items with very little effort. Check out the soya protein section; bags of good soya go a long way in sauce for pasta or rice. *Do not include frozen processed foods here*; they go in the rear end of the trolley, under the handlebar with what will be a small collection of other processed foods. (As for chips (French fries), I suggest you skip the big bags of pre-cut potatoes; pound for pound they are expensive. The chips you make at home from fresh potatoes taste better and cost less. Learn to cut chips by hand or invest in a chip-cutter that cuts 12–18 chips at a time, depending on size.)

Now look over the canned fruit and vegetable sections of the store. As you buy canned foods, place those containing only healthy fruit and vegetable choices next to the frozen foods. Canned tomatoes and corn go here, as do cans of beans and baked beans; all of these are good value for money. So are canned fruits. Like canned vegetables, fruit loses some of its nutrients during the canning process, but it makes a healthier – and often cheaper – dessert than ice cream or cake. A modest lashing of cream on a dish of prunes canned in fruit juice makes a delicious dessert! Keep a grip on your purse strings, however, and avoid cans with fancy names and 'designer' labels. Go for items that are simple. One ingredient per can is a good rule.

Finally, on top of the cans, stack the bread (wholemeal), dried beans and pulses, and wholegrain products: pasta, noodles, dried beans and lentils go here also. This group of foods should take up a major part of your shopping trolley. Leave the cakes and other goodies until last, so you can keep a close eye on how much they are going to cost!

While you are shopping, place other items you buy at the near end of the trolley; any dairy products, meat or fish, plus the eggs, butter and oil also go here. (Skip the expensive bottles of salad dressing and salad cream altogether; it is far cheaper – and far more tasty – to make your own.)

Household and beauty items like bleach, detergent and shampoo should be placed in a sack and hung from a hook on the back of the trolley, or placed underneath on a shelf, if one is provided. This reduces the possibility of contaminating foods though leakage or odour.

Just before heading for the check-out counter, take a good look at the contents of your trolley and see how well you have chosen. The food items at the far end and centre should make up the bulk of your purchases. Bread, wholegrain items and pulses should form the next largest group of purchases. Finally, the meat and fish, milk, eggs, oils and any processed foods should take up less than a quarter of the total volume of items purchased. Ignoring any packaging, is your trolley full of colour? If so, you have probably chosen well.

When choosing fresh fruits and vegetables:
* Buy fresh, organic foods when possible. (If space, weather and time permit, grow your own salad greens and tomatoes.)

TIPS ON BUYING AND USING FOODS FROM PLANTS

When organic foods are not available, pick the best you can find from what is on offer.

✳ Look for two kinds of fresh produce when you shop.

✳ First, look for those items you are going to serve fresh and crunchy within the next 48 hours. Fresh produce loses nutrients when standing around and most of what you see on the shelves has probably travelled farther than you do on your annual holiday! Buy fresh, crisp, dark green and red foods. If a cucumber feels soft and looks tired, put it back; it will do nothing for your evening salad. The spinach looks limp? Pass it by; the vitamin C content has probably already taken a nose-dive. Buy firm apples and pears and allow them to ripen a bit at home as necessary. Anything with a blemish or bruise is going to spoil quickly. Pay top money only for the best.

Remember the special cases: contrary to almost every other fresh food, bananas and avocados are both best when picked green and allowed to ripen at home. For peeling and eating, look for bananas that are slightly green and have no black or bruised areas. Make sure you pack them with your other purchases so that they are not crushed on the way home. Keep them on top and treat with care. Avocados should be bought several days in advance and allowed to ripen at room temperature. Select fruit that is firm to the touch all over and watch out for those that are slightly soft at the top; this may mean they have been bruised. (If you want to speed the ripening process, place avocados in a paper bag and leave in a warm place.)

✳ The second type of food you should shop for are those items you plan to use in a soup, stew or baked product. This means looking for bargains. In some stores there are mark-down labels on items that have reached their sell-by date (the use-by date is usually at least a day later). In other shops, a special counter is set aside for these items, plus any on sale because they are slightly bruised or damaged in some way. Check them over. Anything you can use? Bruised bananas make excellent (and highly nutritious) banana-nut bread; wilted celery is good in soup and that tired spinach could be part of a tasty quiche.

Healthy eating need not be expensive. By cutting down, or eliminating, costly meat in meal planning, and substituting soya products, milk, cheese and eggs wisely, you can serve your family delicious food containing less fat and more nutrients. As a bonus, you may find the range of dishes you prepare is wider, more interesting to eat and less likely to add unwanted inches to family waistlines.

GLOSSARY

Words *in italics* are listed elsewhere in the glossary.

acid-base balance: Normal body processes depend on a balanced chemical environment. Too much acid in the body fluids (acidosis) may occur as a symptom of untreated diabetes; an excess of alkaline chemicals (alkalosis) can be caused by severe and extended vomiting.

aflatoxins: Toxic substances produced by certain types of mould likely to contaminate stored groundnuts (peanuts) and *grain*s, and known to cause liver *cancer* in farm and experimental animals. High concentrations have been found in the diets of people living in hot, humid regions of the world where high rates of liver cancer are recorded.

algae: The simplest members of the plant kingdom (containing about 25,000 species), these single-celled organisms contain *chlorophyll* but lack *stem*s, *roots* and *leaves*. Rich in omega-3 fatty acids, algae provide fish the substrate needed for the formation of *essential fatty acids* we specifically associate with fish *oil*: EPA (eicosapentaenoic acid) and DHA (docosahexaenoic acid)

amino acids: The smallest chemical units joined together to form *proteins* needed for *cell* structure, mobility, normal growth, *metabolism*, reproduction and tissue repair. Under normal conditions, most of the specific amino acids needed for good health can be manufactured by the human body. However, there are eight amino acids that cannot be manufactured in the body and must be obtained from food. In adults, the 'essential' amino acids are: isoleucine, *lysine*, leucine, methionine, phenylalanine, threonine, valine and tryptophan. Rapid growth in children often makes it impossible for the body to keep up with the demand for another amino acid – histidine – making this an 'essential' amino acid in those who have not reached their adult size.

The most concentrated sources of essential amino acids are meat, milk and eggs. *Vegetarian*s, vegans and anyone else who may go for periods of time avoiding animal protein in their food should make certain their diet includes meals providing all of the essential amino acids; this can be achieved by eating soya (soy) beans and products made from soya protein, or by enjoying foods containing a mix of *grain* and beans or peas.

anabolism: The metabolic, or chemical, processes in the body which change simple substances into complex ones; for example, processes that build up complex *protein*s from single *amino acids*.

anthocyanins: Natural chemicals found in blackcurrants, cranberries and other red/black-skinned *fruit*s which inhibit bacterial growth. Eating foods rich in anthocyanins may help prevent or control urinary tract infection. Scientific research suggests anthocyanins help strengthen collagen, like that found in skin. Freezing reduces the amount of these substances in foods.

antioxidants: Chemicals which block the destructive action of *free radicals* and the processes of oxidation.

associate *nutrients*: Numerous substances occurring naturally in plants, which are not required by the human body for any of the processes of life (growth, reproduction, *metabolism* and tissue repair), but which have an effect on these processes. For example, *phytoestrogens*, substances found in soya and other plants, are so similar in structure to the oestrogen produced in humans they can be mistaken for the real thing by the body's *cell*s. Some scientific studies suggest that a high intake of foods rich in phytoestrogens (like that enjoyed as part of the traditional Japanese diet) is linked with a reduced risk of breast *cancer*. We are in the early stages of understanding how these and other associate nutrients influence human health.

atherosclerosis: A degenerative disease in which the arteries feeding the heart muscle gradually become lined, or blocked, with a rough layer of *plaque*, a fatty substance, thereby slowly restricting the normal flow of blood and causing tissue destruction and death.

baking: The processes of cooking by surrounding food with very hot air, as in an oven.

BBI: *See Bowman-Birk Inhibitor.*

beri-beri: A disease cause by a deficiency of *vitamin* B1; common among people living on a diet of polished rice. Symptoms include oedema, nerve degeneration and heart disease.

biotechnology: The theory and methods applied in the alteration of genetic material *(see Genetic Engineering)*.

Bowman-Birk Inhibitor (BBI): A compound found in soya which blocks the action of *enzymes* known to promote tumour growth.

***Brassica*:** A genus of plants including broccoli, Brussels *sprout*s, cabbage, cauliflower, kale and turnips and known to contain chemicals that fight certain forms of *cancer (see also Cruciferae)*.

broiling: Preparing food by placing it over or under a hot grill, thus cooking one side at a time. Similar to *grilling*; however, broiling is usually done in an oven or other enclosed cooking device.

bud: Part of a plant containing the early stages of developing *flower*s, *stem*s, *leaves* and *roots*. Buds are rich in cytokins, which are powerful plant hormones regulating growth. The area of the plant around them is rich in nucleic acids and other substances needed to support rapid growth.

Capers are the most common form of bud used in Western cooking. Although these flavourful condiments are traditionally prepared from the buds of shrubs native to eastern Asia, flower buds from broom, nasturtium, marigold and buttercup may also be pickled and enjoyed as lively additions to sauces and as garnishes.

bulb: The upright underground *stem* of a plant that stores food energy between its seasons of growth. *Roots* grow from the base of a bulb, and the upper stem, *leaves* and *flower*s emerge from the top portion. Garlic and onions are edible bulbs; the *germ*, or dormant embryo of a new plant, is at the core of dense fleshy leaves, or plates, of stored food.

CAD: *See Coronary Artery Disease.*

cancer: An abnormal growth of *cell*s which can spread to other sites (malignant tumour). The causes of cancer are complex and poorly understood; however, certain food choices are linked with high rates of specific cancers. For example, women known to consume diets containing high levels of saturated *fat*s over a long period of time appear to be at greater risk from breast cancer.

cancer initiator: A substance or external effect (excessive exposure to sunlight is an example) capable of causing changes in *cell*s leading to abnormal growth and the development of a cancerous tumour.

cancer promoter: A substance or condition which 'promotes', or encourages, tumour *cell* growth.

carbohydrates: A form of energy manufactured by and stored in plants, and needed by animals for energy and growth. Along with *fat*s and *protein*s, these are major *nutrients* needed for life. They are an important group of chemicals in living things which includes *sugars*, *starches* and *cellulose*. Through the processes of *photosynthesis*, plants combine light energy with carbon, hydrogen and oxygen into simple sugar molecules (saccharides). These are then combined into larger, more complex molecules of starches and *fibre*.

carbon dioxide: A colourless, odourless gas formed as a waste product during the processes of *respiration* in the body. Released into the air through the lungs, this gas becomes part of the atmosphere, from which it is absorbed and used by plants during the process of *photosynthesis*.

carcinogen: Any substance or condition which starts or supports a cancerous growth.

catabolism: A series of chemical reactions in the body which break down substances from the foods we eat into simpler units; *proteins*, for example, are broken down into *amino acids*. Energy is released during catabolism for use in making new substances, *cell* growth, muscle contractions involved in motion and all of the other processes of life.

cataracts: A non-reversible clouding of the lens of the eye, often associated with ageing, which leads to eventual blindness. Excessive *free radicals* are thought to be a major cause and studies have suggested high dietary levels of beta-carotene, a powerful antioxidant, may prevent or slow the progress of this debilitating condition.

cell: The smallest unit in a living organism. Cell growth and differentiation are governed by DNA. Each cell has a defining membrane; in higher plants, this membrane is surrounded by a tougher cell wall containing *cellulose*.

cellulose: A form of indigestible *carbohydrate*, or polysaccharide; the primary source of *fibre* in food.

chelator: Any substance which combines with and holds metals, thus preventing their absorption into the body *(see also Phytic Acid)*.

chlorophyll: The green pigment in plants which is necessary for the conversion of light, water and *carbon dioxide* into food energy. This process is called *photosynthesis*. Chlorophyll is produced synthetically for use as a powerful external deodorant and thought to be a powerful anti-*cancer* agent.

cholesterol: A fatty substance naturally found in animal tissues and fluids. It is used by the human body to form bile salts, needed to digest dietary *fats*, and as a building block in *cell* membranes and steroid hormones. Cholesterol is absorbed from food and manufactured by the liver.

LDL cholesterol: When found in high concentration in the blood, Low Density Lipid cholesterol, also known as 'bad' cholesterol, is the form of this fat most closely linked with an increased risk of *coronary artery disease*. Diets high in saturated fats have been shown to increase the level of LDL cholesterol.

HDL cholesterol: High Density Lipid cholesterol, also known as 'good' cholesterol, is thought to lower the risk of heart disease by minimalizing the amount of fatty *plaque* deposited on arterial walls. Diets rich in *essential fatty acids* have been shown to increase the proportion of HDL cholesterol circulating in the blood.

cobalamin: *Vitamin* B_{12}. Plants are an excellent source of vitamins with the exception of B_{12}, which is found in animal products, especially liver. Spirulina and Chlorella contain cobalamin. Some food spreads made from *yeast* contain B_{12}, but these products probably have been fortified. Good sources for *vegetarian*s and vegans include fortified cereals.

coenzymes: Substances that work together with *vitamins* and *enzymes* in specific biochemical reactions.

coronary artery disease (CAD): Damage to the heart muscle caused by a reduced flow of blood and *nutrients*; usually caused by the build-up of fatty *plaque* on the walls of the arteries which feed the heart *(see also Atherosclerosis)*. Coronary artery disease in the United States is a major cause of death, but its rate has been dropping in recent years. In the United

Kingdom and Ireland, rates remain high.

Cruciferae: A family of about 1,900 plants which are characterized by *flower*s with four petals in the shape of a cross. Widely grown in temperate climates and especially in Europe. These plants contain *phytochemicals* believed to prevent *cancer* and damage to the genetic material in *cells (see also Brassica).*

cystitis: Inflammation of the urinary bladder.

daidzein: An anti-*cancer* compound identified in soya (soy).

deficiency syndrome: Any single or combination of abnormal physical symptoms recognized as attributable to the absence of a specific *nutrient.* For example, *vitamin* C deficiency, known as scurvy, results in a wide range of symptoms including bleeding gums. One or all symptoms may be present in a case of nutritional deficiency, sometimes making diagnosis difficult.

deoxyribonucleic acid: *See DNA.*

diuretic: Chemical substance that stimulates the excretion of water from the body as urine.

DNA (deoxyribonucleic acid): The double-helix molecule, primarily found in the nucleus of *cells* but also found in *mitochondria,* containing the genetic code or 'control pattern' that determines the unique set of characteristics in each living thing.

enzymes: Small *protein* molecules, containing a highly specific series of linked *amino acids,* which control chemical process in the body. Thousands of enzymes participate in normal human *cell* functions. Because essential amino acids are a necessary part of many important enzymes, a diet must provide adequate quantities of 'complete' protein to ensure adequate quantities of essential amino acids are available.

equole: compound formed from *isoflavones* by normal bacteria in the human gut of people eating soya (soy) and soya products.

Equole has oestrogenic properties.

essential amino acids: *See Amino Acids.*

essential fatty acids: Specific molecules of *fat* needed for normal growth and good health which cannot be manufactured in the human body. The two most important essential fatty acids are linoleic acid, chemically known as an omega-6 fatty acids, and alpha-linolenic acid, an omega-3 fatty acid. Scientific research indicates we need a balanced blend of omega-6 and omega-3 fatty acids to prevent degenerative illnesses such as heart disease. Nuts, *seed*s and *oil*y fish are good sources of essential fatty acids. We should consume about five times more omega-6 than omega-3 fats. Both types require adequate levels of *vitamin* E and C to protect their delicate double bond structure.

fat: A greasy substance found in animals and some parts of plants (*seed*s, nuts and certain *fruit*s). Fats are a rich source of stored energy and by weight contain about twice as many calories as either *protein* or *carbohydrate*. They're necessary for the absorption of fat-soluble *vitamins* A, D, E and K. *Cholesterol*, lipoprotein, *triglyceride*s and phospholipid are all types of fat found in a healthy body.
  monounsaturated fat: Fatty acids containing only one double bond. Rich plant sources are avocado and olives. Unlike saturated fats, monounsaturated fats appear to benefit the heart by not competing for the *enzymes* required for *metabolism* of *essential fatty acids* by tissue *cell*s.
  polyunsaturated fat: Long-chain fatty acids in which double bonds exist between two or more carbons in its carbon chain. All essential fatty acids are polyunsaturated; therefore, a reasonable amount of polyunsaturated fats should be included in the diet. It is important to remember that the double bonds in polyunsaturated fats are delicate and can be destroyed by *free radicals*, or oxidation. By ensuring your diet includes good sources of vitamins C and E, harmful oxidation of the double bonds can be controlled.
  saturated fat: Fatty acids containing no unsaturated double

bonds in their carbon chain. Saturated fats may be short or long chains; some research suggests short chain saturated fats are more likely to be dangerous for the heart than the longer varieties. Saturated fats are thought to be one factor that inhibit the activity of the *enzyme* delta-6-desaturase, needed in the metabolism of essential fatty acids. For this reason, some scientists believe diets high in saturated fats should be supplemented with foods or supplements rich in GLA (gamma-linolenic acid, an omega-6 fatty acid) and other metabolites of essential fatty acids.

fat-soluble vitamins: *See Vitamins.*

favism: An inherited sensitivity to a specific substance present in the broad bean *(Vicia faba)* which results in anaemia caused by destruction of red blood *cell*s. The condition is almost always found among people living in the Mediterranean region.

fibre: Fibre is that part of food from plants that cannot be broken down during digestion. It forms the bulk needed for healthy excretion of solid waste from the bowel. There are two types of fibre:

soluble fibre: found in oats and other *seed*s, which 'bulks up' in water. Research suggests this form of fibre helps reduce blood *cholesterol* levels.

insoluble fibre: found in all *fruit*, vegetables, *grain*s, nuts and seeds; which is important for the normal passage of waste from the digestive system.

flatulence: Wind produced in the bowel. *Pulses*, particularly beans, are known to cause flatulence. Adding certain spices and herbs to food helps prevent this problem; these include cumin, bay and ginger.

flour: Finely ground *grain* used in the preparation of food. Wheat flour is the most common in the Western diet, but others are made from rye, barley, rice, corn (maize) and oats. The origins of flour milling trace back to prehistoric times, but these were course products. Roman chefs, with their interest in culinary refinement, required something better, and it was for them that the rotary motion mill was developed.

flower: The reproductive part of a plant from which, after fertilization, the *fruit*s and *seed*s will develop. The colour and scent of the flower, or blossom, attracts the insects and other creatures needed for fertilization.

folate: Another name for folic acid *(see Chapter 2, p.39)*.

free radicals: Molecules which, for a very brief period of time, are missing an electron are therefore highly unstable. They 'seek' a replacement electron from other, stable molecules which are then damaged when attacked by the free radical. The formation of free radicals can occur in a chain reaction. Some free radicals are part of normal chemical processes in the human body; others result from damaging outside factors, including excessive exposure to sunlight, cigarette smoke and atmospheric pollution.
   Research links the formation of abnormally high levels of free radicals with tissue changes leading to *cancer*, heart disease, wrinkled, dry skin and *cataracts*.

fruit: The fleshy, enlarged reproductive part of a *seed* plant, the part which develops from a fertilized *flower* and contains the seeds. Fruits come in many forms and not only include sweet-tasting apples, mangoes and pears, but also cucumbers, tomatoes and pumpkins.

frying: A method of cooking in which food is exposed to heat through a layer of *fat*. The high temperature of the fat evenly heats the surface of a food, evaporates the surface water and changes the structure of the starches and *protein*s. When food is 'shallow fried', only a part of the surface is exposed to intense heat at a time; 'deep frying' is the total immersion of food in hot fat. Draining food well after frying helps remove excess fat and the calories it adds. Always use fresh *oil* when frying.

genetic engineering: Methods of manipulating the normal

patterns of DNA in living organisms to produce a permanent, inheritable change. The DNA may be altered by adding genetic material from another species or causing structural alterations in the normal genetic structure of a species. Long-life tomatoes are an example of a biologically-engineered plant. By adding genetic material from another plant to the normal genetic material in a tomato, scientists were able to produce a tomato which ripened at a slower rate. This new characteristic helped food producers by increasing available shipping time and increased the shelf-life of an otherwise delicate food.

genistein: A substance (*isoflavone*) found in soya (soy) which has been demonstrated to inhibit *cell* growth in both breast and prostate *cancer*. Some scientific studies suggest genistein can return cancerous cells to normal.

germ: That part of a nut or *seed* which contains the embryo of a new plant. This structure is rich in *nutrients*. It remains dormant until the right combination of moisture and temperature stimulate certain internal *enzymes* which begin *cell* division and differentiation; these processes lead to development of the stalk, *roots* and *leaves* of a new plant.

germination: The first stage of plant growth in which the *fats* and starches in seeds are changed into *amino acids* (*proteins*), *vitamins* and other substances needed for development. Moisture and warmth are needed to start germination.

glucosinolate: A classification of *cancer*-fighting chemicals found in cruciferous vegetables.

grain: The highly nutritious *seeds* of cereal grasses, such as wheat, barley, rye, corn (maize) and oats. Grain consists of three parts: the tough outer cover, called bran; a small, nutritionally dense segment, the *germ*, from which a new plant will form; and a large energy rich store of *carbohydrate* and *fats*, known as the endosperm. The first steps in the milling process separate the endosperm from the germ and bran, after which it is cleaned, bleached and milled to one of several grades of refinement.

THE PLANTS WE NEED TO EAT

gram: A metric measure of weight (28.35 grams = 1 ounce)

grilling: Cooking by exposing food, one side at a time, to a hot, dry surface. Grilled food is healthier than fried because little or no *fat* is required for cooking.

hypertension: Increased blood pressure in the main arteries which exists even when the person is at rest. Unless well advanced, this is a silent disease which, in about 90 per cent of cases, has no underlying cause. Risk factors associated with hypertension are sex (males are more susceptible than females), family history, being overweight, excessive alcohol consumption, excessive consumption of salt (sodium) and smoking.

hypertensive crisis: Sudden onset of dangerously high blood pressure, or *hypertension*.

indoles: Substances in plants (*phytochemicals*) which are thought to offer some protection against *carcinogen*s, especially those that cause breast *cancer*. They are thought to work by blocking the activity of natural oestrogens. Broccoli, kohlrabi and other cruciferous plants are good sources of indoles.

inorganic: Pertaining to chemicals not associated with the processes of life; chemicals which do not contain the element carbon.

isoflavones: Compounds found in plants which mimic the effect of low doses of oestrogen in the human body. Research suggests these compounds help protect against *cancer* and can also reduce the troublesome symptoms of the menopause. *Indoles* are thought to act on the liver, where they stimulate the breakdown of excess oestrogen; this may aid in preventing breast cancer.

isolated soya *proteins* (ISP): A commercial form of soya which contains not less that 90 per cent protein.

isothiocyanates: Plant chemicals believed to block the activity of carcinogenic substances that affect the lower bowel. Found in cruciferous plants.

ISP: *See Isolated Soya Products.*

IU: International Unit. An international system of measurement, based on the metric system, adopted in 1960 by the International Organization for Standardization; however, *mg* and IU Recommended Daily Allowances (RDAs) are not always the same for *vitamin* activity and care should be taken not to confuse the two.

leaf: The thin, expanded extensions of *stem*s rich in *chlorophyll* and specially adapted to capture and combine the energy from sunlight with water and *carbon dioxide* to produce *carbohydrate* molecules

lecithin: A specific phospholipid (*fat*) known to emulsify fats. Contains choline and inositol.

legume: In France and other parts of Europe legume is often used to describe plants used as vegetables. However, the true meaning of the word has to do with the structure of its *fruit*: peas, beans and other *seed*s that form on only one side of a pod, or carpel, are true legumes. These seeds have a high *carbohydrate* content and are often dried for use months and even years later.

lipids: A group of energy-rich components found in all living things. They are insoluble in water, but dissolve in *organic* solvents such as ether and chloroform. Organic chemicals known as lipids include *cholesterol*, steroid hormones, *triglyceride*s (the most abundant form of fat found in living tissues), phospholipids (complex fats incorporating the mineral phosphate) and waxes.

Lupus erythmatosis: A chronic, autoimmune disease of connective tissue which may be influenced by diet.

lysine: An essential *amino acid* missing from most plant *protein*s, but plentiful in animals. Soya (soy) and *yeast*-based products are lysine sources acceptable to vegans; however, these foods must be consumed with a *grain* product in order for the lysine to be fully absorbed.

MAOIs (monoamine oxidase inhibitors): A type of anti-depressant drug.

mcg: Microgram; one millionth of a *gram*.

Mediterranean diet: A combination of foods normally enjoyed by people living in and around the coastal regions of the Mediterranean Sea, where epidemiological studies have shown rates of heart disease and *cancer* to be lower than those found in most populations living on high-*fat* Western diets. The Mediterranean diet includes seafood, but is primarily based on foods derived from plants, including *oil*s, nuts, fresh *fruit* and vegetables, and *grain*s. Tomatoes and garlic – two ubiquitous ingredients in Mediterranean cooking – are rich sources of natural *antioxidants* and other healthy *phytochemicals*.

metabolism: Chemical processes in the body which either build up substances needed for reproduction, growth, motion and tissue repair (*anabolism*), or break down *cell* substances (*catabolism*) to release energy or perform specific functions.

methylmethionine: Research suggests a natural chemical found in cabbage, S-methylmethionine, supports healing of stomach ulcers and suppresses tumour formation in the lower bowel.

mg: Milligram; one thousandth of a *gram*.

microwave cooking: A means of cooking food by exposure to a source of high-frequency, ultra-shortwave energy. Energy waves excite molecules on and near the surface of the food, creating heat which is then passed deep inside. For this reason, most foods cooked in a microwave oven should be stirred at least once during cooking. Considerable research remains to be done concerning the effect of microwaves on *nutrients*; however, there is evidence it destroys less of the *vitamins* than boiling or steaming.

miso: A fermented soya (soy) bean paste.

mitochondria: Small, easily identifiable and well-formed sub-structures within living *cells*, responsible for the release of energy from *nutrients*.

monoamine oxidase inhibitors: *See MAOIs.*

nightshade: Plants of the genus *Solanum* which include some of our favourite vegetables: tomatoes, aubergine (eggplant), potatoes and box (bell) peppers. These plants have a distinctive *flower* and may cause skin reactions. The name is usually associated with another member of the group, the herb deadly nightshade, or *Belladonna*, long known to contain atropine, which may be used in a tincture to treat asthma, colic and hyperactivity.

nitrites: Salts contain nitrogen which are used to preserve meat. Often used with potassium nitrate (saltpetre) to stop bacterial growth. Nitrites are converted into nitrosamines in the body; in high quantities these have been shown to cause *cancer* in animals, but there is little clear evidence they have the same effect in humans.

nutrients: Substances in food needed by the body to carry out the specific chemical processes of normal growth and good health; these include *proteins*, *carbohydrates*, *fats*, *vitamins* and minerals. Water is also a nutrient and is one of the most vital elements in maintaining good health. *See also Associate Nutrients.*

obesity: A condition in which the body contains an excessive amount of *fat* as compared with average members of the population. Obesity results from an intake of food energy that exceeds the energy used during the course of activity; excess energy is stored as *triglycerides* in fat *cells*. Although obesity has been statistically linked with increased rates of heart disease, diabetes and certain forms of *cancer*, there is no proof that obesity alone causes these diseases.

oil: A fluid form of *fat*. Oils consist almost entirely of *triglycerides* and are, therefore, very high in calories. Oils are excellent sources of polyunsaturated fatty acids, including those known as *essential fatty acids*.

organic: Chemical substances derived from living organisms; a class of substances containing carbon.

oxalic acid: A natural substance in spinach and rhubarb that blocks the body's ability to absorb calcium and iron. These foods should not be included in the same meal as food eaten to supply these minerals.

pectin: A form of soluble *fibre* that is thought to reduce blood *cholesterol* levels.

pesticides: Chemicals that kill insects. Some plants produce their own pesticides in the form of *terpins*.

photosynthesis: The chemical processes by which green plants convert *carbon dioxide*, water and the energy from sunlight into starches (*carbohydrates*). *Chlorophyll* is an integral part of this process. During the process of photosynthesis, carbon dioxide is removed from the air and returned into the atmosphere as oxygen.

phytic acid: A substance found in plants, especially soya (soy), which binds iron and zinc in food and decreases their absorption into the body.

phytochemicals: Biologically active compounds found in plants; many have been shown to have healing properties in humans. Examples are daidzein (found in soya), carotenoids (the coloured substances in red and yellow plants) and allium compounds (the stuff that gives garlic its zing!)

phytoestrogens: *Phytochemicals* with oestrogen-like effects in the human body.

protein: One of the major macro-*nutrients* needed for normal growth and health, proteins are the primary source of nitrogen in the diet. Proteins are complex substances constructed from very specific sequences of *amino acids*. If any of these building blocks are unavailable, errors may occur in body chemical functions because the correct protein form is absent. Animal proteins contain all of the 20 amino acids human growth and well-being require. In plant protein one or more essential amino acids may be missing. If no animal protein is included in a diet, it is vitally important to include a number of foods which combine *grain*s and *seed*s or *pulses* to ensure inclusion of the total mix of essential amino acids.

pulses: In the context of foods, edible *seed*s from leguminous plants *(see Legume)*. These include chickpeas, peas, beans, broad beans, soya, lentils and peanuts. Pulses are a rich source of *carbohydrate*, minerals and *protein*. However, because their protein lacks certain essential *amino acids*, needed by processes in the human body, *vegetarian*s and vegans should always combine pulses with rice, bread, or other *grain* products, to ensure a complete and healthy diet.

purines: a group of substances produced in the body during the breakdown, or digestion, of certain *protein*s. Abnormally high levels of purines can lead to a raised level of uric acid (hyper-uricaemia) which may cause *gout*, a metabolic disorder that causes attacks of inflammation and pain in single joints, particularly those at the base of the big toe, the ankle, knee and wrist. Gout may also cause the formation of deposits in the kidney which may result in kidney failure. Purines are found in *pulses*, foods containing caffeine and certain meats (poultry, liver and kidney).

RDA: This usually means 'Recommended Daily Allowance' but can mean 'Recommended Daily Amount', but these two measures are not equivalent. Both are quantities of specific *vitamins* and minerals thought to be adequate for normal growth and good health. Some nutritionists are now questioning the validity of publishing RDAs because they reflect a norm in the population which may not be adequate or appropriate to meet the age and health requirements of an individual. RDAs published by different health organizations do not always agree.

respiration: All of the chemical and physical processes involved in an organism using oxygen from the air and eliminating the *carbon dioxide* which results as a waste product.

roasting: A cooking method using dry heat, in which food is kept moist either with juices exuded during the cooking process or a mixture of *oil*s and flavourings. Basting or brushing foods with oil before and during roasting can improve their texture and flavour by intensifying the surface heat and reducing moisture loss.

roots: Those parts of a plant anchoring it into the ground and absorbing water and mineral *nutrients* from the soil around it. Some roots (potatoes, Jerusalem artichokes) produce edible, nutritious *tuber*s. Others, such as carrots, store *carbohydrate* and other nutrients around the central core of the root. Onions and garlic produce bulbous *stem*s below the ground which contain the *germ* of a new plant; at the base of the swelling roots expand out in an unorganized fashion *(see also Bulb)*.

seed: *Grain* or other ripe, fertilized ovule, or reproductive organ, of a plant. Along with the *germ*, or embryonic new plant, seeds contain a rich mixture of *carbohydrates*, *fats*, *vitamins* and minerals to support new growth when it begins. The balance and mix of these *nutrients* is a characteristic of the plant. Low in water content, seeds (which include nuts) are good sources of highly concentrated nutrition. Seeds are encased in a protective covering which adds to the roughage in the diet.

sprout: The early growth from a *bud* or *seed*. Stimulated by hormonal changes, which may be caused by changes in temperature or available moisture, *cells* in a *germ*, or embryonic plant, begin to multiply and form tender shoots which are far higher in water content and lower in *fat* and *carbohydrate* than the seed or section of the plant from which they grew. Sprouts from certain beans (aduki, mung) are delicious additions to salads, adding mineral and flavour content to a dish.

stem: The main aerial axis of a plant which supports the *leaves*, *flowers* and *fruit* and serves as the conduit through which water and *nutrients* pass between its various parts.

sugars: Sweet tasting, simple *carbohydrates*. Examples are glucose (the kinds of carbohydrate used by body *cells* for energy), lactose (sugar found in milk) and sucrose (table sugar). Most forms of sugar are derived from plants.

synergist: A substance that works to enhance the effects of another substance.

tempeh: A meaty-flavoured soya (soy) product, popular in Indonesian foods.

terpins: Small *lipid* molecules formed only by plants which may be antibacterial in nature, natural insecticide; one form of essential *oils*. Terpins in pine are a natural insecticide and produce the trees' characteristic odour.

tofu: Soya (soy) bean curd, popular in most of Asia.

toxins: Poisonous substances naturally produced by living organisms.

triglyceride: The most common form of *fat* found in the body, consisting of one molecule of glycerol and up to three molecules of fatty acids. The fatty acids may be saturated, monounsaturated, polyunsaturated, or a mix of the three. High blood triglyceride levels may be a good indicator of *coronary artery disease*.

tuber: A thickened part of a plant's underground *stem* or *root* in which *carbohydrates* and other *nutrients* are stored. 'Eyes', or *germination* points, from which new plants can develop are to be seen on the surface. Potatoes are examples of edible tubers.

umbelliferous: Herbaceous shrubs and plants characterized by hollow *stem*s, divided compound *leaves* and *flower*s having stalks of the same length that grow in clusters arising from single points on the main stem. Parsley, fennel, parsnips and carrots are examples.

vegetarian: It is sometimes difficult to know what people mean when they claim to be vegetarians. Most health-conscious diners choose to be 'demi-vegetarians': eating no red meat and only limited amounts of fish and poultry. True vegetarians eat no food that involves killing animals, including fish and other forms of seafood. Eggs, milk and cheese are permissible, however, and represent good sources of total, or complete, *protein*.

Vegans completely avoid all foods with an animal origin, including milk, eggs, honey and royal jelly. Vegans are in danger of *vitamin* B_{12} (*cobalamin*) and vitamin D deficiency. The first problem can be overcome by including Spirulina or fortified foods in the diet, and the body will produce its own vitamin D when the skin is exposed to moderate amounts of sunlight. Vegan diets take planning to ensure nutritional balance in adults.

vitamins: Substances which cannot be manufactured by the human body, but which are indispensable for human health and well-being. Unlike the macro-*nutrients*, *protein*s, *carbohydrates* and *fat*s, very small amounts of vitamins are required in the diet. For decades, medical experts have prescribed specific amounts of vitamins, RDAs, for the general public. Recent research, however, suggests these quantities may be low in certain cases. Also, because people vary in both their nutritional and physical condition, specific vitamin requirements may vary widely between individuals.

There are two types of vitamins: fat-soluble and water-soluble. Fat-soluble vitamins (vitamins A, D, E and K) are absorbed from food when fat is present; they can be stored in body tissues. In excessive amounts, vitamin A may become toxic. Water-soluble vitamins (vitamin C and B-complex vitamins) are not stored in the body and do not require dietary fat for absorption. *See Chapter 2 for more information about individual vitamins.*

yeast: Single-celled fungus used in brewing and fermenting foods. Yeast extract is used as food because of its high *vitamin* C content.

SOURCES AND
SUGGESTED READING

Avery Jones, Sir Francis, 'Review: new concepts in human nutrition in the twentieth century: the special role of micro-nutrients', *Journal of Nutritional Medicine*, 4 (1994), pp.99–113

Banks, Tony, 'Vegetarianism', *Hansard*, 8 March 1995, pp.344–8

Bissell, Frances, *Sainsbury's Book of Food*, Websters International Publishers, London, 1989

Bland, Jeffrey S., 'Oxidants and antioxidants in clinical medicine: past, present and future potential', *Journal of Nutritional and Environmental Medicine* 5 (1995) no.3, pp.255–80

Blythman, Joanna, *The Food We Eat*, Michael Joseph Ltd, London, 1996

Campion, Kitty, *Kitty Campion's Vegetarian Encyclopedia*, Random House, London, 1995

Castelvetro, Giacomo, *The Fruit, Herbs & Vegetables of Italy*, Viking, London, 1989

Chaitow, Leon, *Stress: Proven Stress-Coping Strategies for Better Health*, Thorsons, London, 1995

Coultate, T. P., *Food: The Chemistry of its Components*, The Royal Society of Chemistry, Cambridge, 1989

Coultate Tom and Davies, Jill, *Food: The Definitive Guide*, The Royal Society of Chemistry, Cambridge, 1994

Crawford, Michael and Marsh, David, *Nutrition and Evolution* Keats Publishing, Inc., Connecticut, 1995)

Davies, Stephen, 'Scientific and ethical foundations of nutritional and environmental medicine. Part II: Further glimpses of "the higher medicine" ', *Journal of Nutritional and Environmental Medicine* 5 (1995), no.1, pp.5–11

Eagle, Robert, *Herbs, Useful Plants*, BBC Books, London, 1981

Erdman, J. W. and Fordyce, E. J., 'Soy protein and the human diet', *American Journal of Clinical Nutrition* 49 (1989), pp.725–37

Ewin, Jeannette, *The Fats We Need to Eat*, Thorsons, London, 1995

'Fats and oils in human nutrition. Report of a joint expert consultation', FAO and WHO nutrition paper, World Health Organization, Italy, 1993

Ferne, A., 'The Great British potato: a study of consumer demand, attitudes and perceptions', *British Food Journal* 94 (1992), no.65, pp.22–8

Foods that Harm – Foods that Heal: An A–Z Guide to Safe and Healthy Eating, Reader's Digest Association Limited, London, 1996

Gale, Mary and Lloyd, Brian, *Sinclair*, The Founders of Modern Nutrition, The McCarrison Society, 25 Tamar Way, Woose Hill, Wokingham, Berkshire, RG11 9UB, UK

Garrow, J. S. and James, W. P. T., eds, *Human Nutrition and Dietetics*, Churchill Livingstone, Edinburgh, 1993

Grigson, Jane, *Jane Grigson's Vegetable Book*, Penguin Books, Harmondsworth, 1980

Hegsted, D. M., 'Nutrition: the changing scene – 1985 W O Atwater memorial lecture', *Nutrition Reviews* 43 (1985) no.12, pp.357–67

Holland, B., Unwin, I. D. and Buss, D. H., *Cereals and Cereal Products: The Third Supplement to McCance & Widdowson's 'The Composition of Foods'* fourth edition, Royal Society of Chemistry, Ministry of Agriculture, Fisheries and Food, London, 1988

––, *Vegetables, Herbs and Spices: The Fifth Supplement to McCance & Widdowson's 'The Composition of Foods'* fourth edition, Royal Society of Chemistry, Ministry of Agriculture, Fisheries and Food, London, 1991

Horrobin, D. F., ed., *Omega-6 Essential Fatty Acids: Pathophysiology and Roles in Clinical Medicine*, Wiley-Liss, New York, 1990

––, 'Nutritional and medical importance of gamma-linolenic acid', *Progress in Lipid Research* 31 (1992) no.2, pp.163–94

Jones, Kenneth, *Shiitake: The Healing Mushroom*, Healing Arts Press, Rochester, Vermont, 1995

Larousse Gastronomique, Hamlyn Publishing Group, London, 1991

Lawson, Harry, *Food Oils and Fats: Technology, Utilization, and Nutrition*, Chapman and Hall, New York, 1995

Maury, E., *Your Good Health: The Medicinal Benefits of Wine Drinking*, Souvenir Press, London, 1992

McCance and Widdowson *The Composition of Foods*, fifth edition, Royal Society of Chemistry and Ministry of Agriculture, Fisheries and Food, London, 1991

'Mediterranean diet: a symposium of the Swiss Society for Nutrition

Research, Lugano, September 30/October 1 1994. Extended summaries of presentations', *International Journal for Vitamin and Nutrition Research* 65 (1995) no.1, pp.56–74

Meyer, Clarence, *American Folk Medicine*, Plume, The American Library, New York, 1973

Mindell, Earl, *Earl Mindell's Food as Medicine*, Fireside, Simon & Schuster, USA, 1994

—, *Earl Mindell's Soy Miracle*, Fireside, Simon & Schuster, USA, 1995

Ody, Penelope, *The Herb Society's Complete Medical Herbal*, Dorling Kindersley, London, 1993

'Proposed nutrient and energy intakes for the European Community: a report of the scientific committee for food of the European Community' *Nutrition Reviews* 51 (1993) no.7, pp.2209–14

Recommended Dietary Allowances, Food and Nutrition Board, National Academy of Sciences, Washington, DC, 1980

Schmid, Ronald F., *Native Nutrition: Eating According to Ancestral Wisdom*, Healing Arts Press, Rochester, Vermont, 1987

Smith, Paul, *Blue-Green Algae*, Thorsons, 1996

Tannahill, Reay, *Food in History*, Stein and Day, New York, 1973

Terrass, Stephen, *Candida Albicans: How your Diet Can Help* Thorsons, London, 1996

Tietze, Harald, *Kombucha – Miracle Fungus: The Essential Handbook*, Gateway Books, Bath, 1994

van Straten, Michael and Griggs, Barbara, *Super Foods*, Dorling Kindersley, London, 1991

Waldron, K. W., Johnson, I. T. and Fenwisk, G. R., eds, *Food and Cancer Prevention: Chemical and Biological Aspects*, The Royal Society of Chemistry, Cambridge, 1993

Walker, Ann, ed., *Applied Human Nutrition for Food Scientists and Home Economists*, Ellis Horwood Limited, London, 1990

INDEX

aflatoxin 206
ageing 53, 78, 183–4
alcohol 50, 57, 58, 60, 61, 62, 66
alfalfa 83
algae (blue-green) 14, 28, 56, 84, 206
allergies 150, 191
amino acids 24, 27–8, 42–4, 206
• *see also* protein
anaemia 185
animal nutrition 27–9
antioxidants 30, 45, 53, 65, 78, 139, 178, 207
• *see also* vitamins C & E, beta-carotene, selenium
apples 84–6
apricot 86
artichokes 86–8
asparagus 88–9
asthma 185
atherosclerosis *see* heart disease
aubergine 89–9
Avery-Jones, Sir Francis 23
avocado 90

banana 91
Banks, Tony (MP) 19–21
beans 92–6
• bean sprouts 31–3, 136, 223
• *see also* pulses
Bedford, Countess of 19
beer, ale and stout 173
beetroot 96
beri-beri 47, 57, 208
bioengineering *see* genetic engineering
biotechnology 208
biotin *see* vitamins
blackberries 96–7
blackcurrants 97–8

bone 50, 52, 69–70, 72, 74, 76
• arthritis 52
• osteoporosis 52, 189
bouquet garni 154, 163, 200
bran 143–4
brassicas 13, 99, 133, 208
bread, pasta, flour 144–5
broccoli 13, 99
Brussels sprouts 99
bud 208–9
Bush, President George 13

cabbage 13, 100–101
calories 37–8, 41
cancer 6, 13, 14, 52, 176–8, 209
• carcinogens 206, 210
candida albicans 141, 185–6
cantaloupe 101–2
cape gooseberries 110–11
carbohydrate 209
carbon dioxide 209
Carleton, Sir Dudley 18
carotenoids 39
carrots 102, 225–6
Castelvetro, Giacomo 18
catabolism 210
cataracts 210
cauliflower 13, 103–4
celeriac 104
celery 104–5
chard 105
Chapman, Jonathan 85
CHD *see* heart disease
cherries 105–6
chicory 106
children 39, 55, 72, 73
• enjoying food 195–8
• *see also* specific nutrients
chilli peppers *see* peppers

THE PLANTS WE NEED TO EAT

chlorella *see* algae
chlorophyll 25–6, 210–11
chocolate 173–4
cholesterol 67, 188, 211
• HDL 211
• LDL 67, 211
cobalamin *see* vitamin B$_{12}$
Columbus, Christopher 8, 107, 137
contraceptive pills 50, 57, 58, 60
cooking methods 87–9, 192–5, 208
• bake 208
• broil 208
• fry 215
• grill 217
• microwave 219–20
• with oil 166–8
• roast 223
corn (maize) *see* grain
Countess of Bedford 19
courgette 107
cranberries 13, 108
Crawford, Dr Michael 1, 17
crucifers 13, 99, 122, 131, 139, 212
cucumber 108–9
Culpeper, Nicholas 162
Cunningham, Marion 196–7
cystitis 13, 212

dandelion 109–10
dates 110
deficiency syndrome 212
• *see also* Chapter 2
digestive problems 57, 186–8
diuretic 41, 212
• *see also* caffeine
DNA and RNA 24, 212
• *see also* genetic engineering

Endicott, John 87
essential fatty acids 15, 84, 213
• *see also* fats
essential nutrients defined 68
exotic fruits 110–11

fatigue 187
fats 2, 17, 28, 44–5, 165–8, 181,
   213–14, 218, 221
favism 214
female reproduction (nutrition) 39,
   60
• *see also* Chapter 2
fibre 30, 45–6, 214
figs 112

flatulence 214
flour 60, 64, 144, 214–15
folic acid 39, 62
• *see also* vitamins – folate
food choices 7, 16–18
food pyramids 36–8
free radicals 215
fruit 215

Galen 47
gallstones (and bile) 59, 187–8
garlic 112–13
genetic engineering 9–10, 215–16
germ *see* plant
germination 216
ginger 113
gluten *see* allergies
Gorinsky, Conrad 157
grain 141–2, 216
• barley 142–3
• buckwheat 145
• bulghar 145–6
• corn, maize (pollenta) 146
• millet 147
• oats 147–8
• rice 148–9
• rye 149
• wheat 150
• wild rice 151
grapefruit 113
grapes 114
grass 26
Grigson, Jane 102, 135

hair *see* skin
health conditions *see* Chapter 2
heart disease 53, 178–81, 211, 224,
   208
herbs, general 152–3
• basil 153–4
• bay 154
• borage 154–5
• chervil 155
• chives 155–6
• coriander 156
• cress 156
• dill 157
• fennel (seeds) 157–8
• marjoram 158
• mint 159
• oregano *see* marjoram
• parsley 159–60, 225
• rosemary 160–61

- sage 161
- savory 161–2
- tarragon 162
- thyme 162–3
human requirements 24, 39–41
- *see also* specific nutrients
hygiene 192–4
hypertension 79, 188, 217

immune system 14, 50, 60, 65, 79, 181–3
isolated soya (soy) protein (ISP) 217
IU (International unit) 218

jams and jelly 97
Jefferson, Thomas 100, 127, 133

kale 13, 114–15
kiwi fruit 115
kohlrabi 115–16
kombucha 116
kumquat 111

lecithin 218
leeks 116–17
legumes and pulses 218
lemons 117
lentils 118
lettuce 118
limes 119
Lind, John 48, 119
lipids *see* fats
logans 111
loquats 111
lupus erythmatosis 219
lychees 111
lysine 219

mango 119
MAOIs 91, 93, 141, 220
Medici, Catherine de 87
Mediterranean diet 8, 21, 138, 153, 163, 219
melons 120
metabolism 207, 219
minerals 27, 67–80
- calcium 69–70
- chloride 70
- chromium 71
- cobalt 71
- copper 71–2
- fluoride 72–3
- iodine 73

- iron 73–4
- magnesium 74–5
- manganese 75
- molybdenum 76
- phosphorous 76–7
- potassium 77
- selenium 54, 77–8
- sodium 79
- zinc 79–80
miso *see* soya
mitochondria 220
molasses 174
mouth ulcers 188
mushrooms 14, 120–21
mustard – greens 122

Napoleon, Bonaparte 137
nectarines 122
nervous system 45, 47, 59–62, 75, 77
- eyes and vision 50, 75
- mental states 57, 61–4
nettles 122–3
Newton, Sir Adam 18
niacin *see* vitamin B₃
nightshade plants 220
nitrites 220
nutrients: definition and origins of 2, 14–16, 24–40, 220
- associate nutrients 14 *see* phyto-chemicals
- good plant sources *see* Chapter 2
- macro-nutrients 41–6, 225
- micro-nutrients 46–80
nuts and seeds – general 163–4
- almonds 164
- Brazil nuts 164
- cashew 164
- hazelnut 164
- peanuts 164
- pecans 164
- pine nuts 165
- pumpkin seeds 165
- sesame seeds 165
- sunflower seeds 165
- walnuts 34, 165

oats *see* grains
obesity *see* weight control
oil – general 165–8, 221
olives 123
onions 124
oranges 124
oxalic acid 221

pantothenic acid *see* vitamins
papaya 125
parsnips 125–6, 225
passion fruit 111
pasta 144
Pauling, Linus 66–7
peaches 126
pears 126–7
peas 127
pellagra 64
peppers 106–7, 127–8
persimmon 111
pesticides 29, 221
photosynthesis 6, 25–6, 221
phytochemicals 5, 12–13, 14, 26,
   38–9, 94–5, 221
• allium 221
• allyl 112
• anthocyanins 98, 207
• associate nutrients 207
• bioflavonoids 38
• brassinosteroids 39
• bromelain 128
• caffeine 50, 57, 58, 60, 173
• capaicin 106
• daidzen 212, 222
• diallyl sulphite (DAS) 112
• ellagic acid 106, 114, 136
• equol 212–13
• genistein 95, 216
• glucosinolates 139, 216
• indoles 217
• isoflavone 217
• isothiocyanates 218
• lentinan 14
• limonene 113, 117
• methol 159
• papin 125
• phenylethylamine 173
• phytic acid 95, 118, 143–4, 149,
   221
• phytoestrogens 207
• purines 89
• quinones 160
• resveratrol 114
• s-methylmethionine 100, 219
• sulphoraphane 99
• terpins 224
• 3-n-butylphthalide 104
• thymol 163
pineapple 128–9
plants, general 28–31
• nutrients in 6, 30 *see* Chapter 2

plants, parts of 81–2
• bulb 209
• flower 65, 81, 215, 225
• fruit 30
• leaf 30, 218, 225
• root 31, 223
• seed 31
• seed germ 216, 224
• stem 30
potatoes 129–30
protein 42–4, 222
prunes 130
pulses (legumes) 22
• mung beans 31–2
• beans:
  • aduki 92
  • black eyed peas 92
  • borlotti 92
  • broad 93
  • cannellini 93
  • chickpeas 93
  • French beans (flagelor) 93–4
  • lima 94
  • mung 31–2, 94
  • navy 94
  • pinto 94
  • red kidney 194
  • soya *see* soya
pumpkin 130–31
pyridoxine *see* vitamin B_6

Quinghaosu 38
Quorn 34–5, 131

radish 131–2
raisins 132
raspberries 132
RDA 21, 40–41, 222, 225
• *see also* specific nutrientsres-
  piration 25, 223
'restless' legs 189
riboflavin *see* vitamins
rickets 52
• *see also* vitamin D

salads 193–5
salt 180–81
scurvy 47, 139
sea kale 133
sea vegetables 133–4
• agag-agag, arame, kelp, laminaria,
  laver, sampire, wakame seeds 223
• sprouted 136

INDEX

• *see also* nuts and seeds
shallots *see* onions
shiitake mushrooms 13, 120–21
• *see also* mushrooms
shopping tips 82–3, 198–205
Sinclair, Hugh M 14–16, 18
skin and hair 39, 50, 52, 59, 60, 62, 64, 65, 75, 79 186
smoking 39, 46, 50, 53, 61, 62, 66
sorrel 134–5
soup (pea) 201
soya 12, 183, 194
• miso 220
• tempeh 224
• tofu 224
spices – general 168
• black pepper 168
• caraway 169
• cinnamon 169
• cumin seeds 169–70
• curry 170
• fenugreek 171
• mustard 171
• nutmeg 171–2
• paprika 172
• saffron 172–3
spina bifida 62, 189
• *see also* folate
spinach 135
spirulina *see* algae
Stauber, Karl 21
stem 224
strawberries 136–7
stress 39, 59, 60
stroke 190
sugar 181, 224
swede 13
sweet potato 137
Szent-Gyorgi, Albert 48

tea, green 13, 174
teeth 13, 50
• *see also* bones
thiamin *see* vitamins
thyroid 73, 191
Tips 99, 127, 140, 143–4
• budget cooking 198–205
• children and food 195–8
• cooking 192–5
• *see also* Facts and Tips in

Chapter 3
tocopherol *see* vitamin E
tomatoes 138
toxins 224
triglycerides 224
tuber 31, 225
turnips 138–9

umbelliferae 126, 158, 225
USA (also Americans) 1, 5, 13, 21, 85, 148

vegan *see* vegetarians
vegetarians 8, 19–20, 27–8, 43, 206–7, 225
vitamins 225
• beta carotene 32, 51
• biotin 64–5
• fat soluble 49–55
• folate (Folic acid) 62–3
• vitamin A (Retinol) 49–51
• vitamin B_1 (Thiamin) 56–7
• vitamin B_2 (Riboflavin) 57–9
• vitamin B_3 (Niacin) 63–4
• vitamin B_5 (Pantothenic acid) 59–60
• vitamin B_6 (Pyridoxine) 60–61
• vitamin B_{12} (Cobalamin) 61–2, 182, 211
• vitamin C 47–8, 65–7
• vitamin D 35, 51–3
• vitamin E 30, 35, 45, 53–5
• vitamin F *see* essential fatty acids
• vitamin K 54–5
• water soluble 55–80

warnings 50, 89, 90, 91, 92, 93, 97, 108, 150, 188, 191
water 41
watercress 139
watermelon 140
weight control 3, 21, 184, 220
wholemeal 23
wine 175
World Health Organization 19, 44

yams 140–41
yeast 29, 56, 141, 226

zucchini *see* courgette

THE PLANTS WE NEED TO EAT